Te Matau a Māui

Fish-hooks, Fishing and Fisheries in New Zealand

Chris Paulin

with Mark Fenwick

University of Hawai'i Press
Honolulu

Published for distribution in New Zealand and the Cook Islands by fishHook Publications
5 Rosetta Road, Raumati South, Paraparaumu 5032, New Zealand.
fishHookpublications@gmail.com   www.fishhookpublications.freewebhost.co.nz
ISBN 978-0-473-32869-6

Published for distribution outside New Zealand and the Cook Islands by University of Hawai'i Press
2840 Kolowalu Street, Honolulu, HI 96822, USA.
www.uhpress.hawaii.edu
ISBN 978-0-8248-6618-1

Library of Congress Cataloging-in-Publication Data

Names: Paulin, C. D., author. | Fenwick, Mark, author.
Title: Te matau a Maui : fish-hooks, fishing and fisheries in New Zealand /
    Chris Paulin with Mark Fenwick.
Description: Honolulu, HI : Published for distribution outside of New Zealand and the Cook Islands by
    University of Hawai'i Press, [2016] | ©2016 | In English with summary in Maori. | "First published in New
    Zealand and the Cook Islands by: fishHook Publications." | Includes bibliographical references and index.
Identifiers: LCCN 2015041322 | ISBN 9780824866181 hardcover
Subjects: LCSH: Fishing–New Zealand–History. | Maori (New Zealand people)–Fishing–History. |
    Fisheries–New Zealand–History. | Fishhooks–New Zealand.
Classification: LCC DU423.F5 P38 2015 | DDC 639.2089/99442–dc23 LC record available at
http://lccn.loc.gov/2015041322

Cover design by Mandy Bayliss. Design, artwork and project management by Streamline Creative Ltd,
Auckland, New Zealand. Typeset in Adobe Garamond Pro and Myriad Pro.
Printed in China by Prolong Press Ltd.

21 20 19 18 17 16   6 5 4 3 2 1

Front cover art: Māori fish-hook: composite wood and bone components, lashed with dressed flax fibres
(muka). Pitt Rivers Museum, Oxford, 1884.11.47 Back cover art: Māori canoe offshore south-west of Mount
Taranaki, with seabirds possibly feeding on discarded offal or bait. Hand-coloured lithograph by G.F. Angus,
1846. Alexander Turnbull Library, Wellington, New Zealand, PUBL-0014-02
Front endpaper: Peter Buck studying Paratene Ngata making an eel basket. Alexander Turnbull Library,
Wellington, New Zealand, 1/2-037930-F (retouched) Back endpaper: A catch of groper, circa 1910.
Photographer unidentified. Alexander Turnbull Library, Wellington, New Zealand, 1/2-021592-G (retouched)

For Kim

*"Each taonga's ancestral pathway has woven a pattern
of human interconnections upon the land for generations,
forming a korowai, or cloak, of knowledge."*

Sir Peter Tapsell (1930-2012)
Te Arawa

*"… The use of marine resources was a fundamental feature
of… mahinga kai, and played a key role not only in the…
economy, but in… social and spiritual life…"*

Ngāi Tahu Sea Fisheries Report,
New Zealand Waitangi Tribunal 1992: 8

Rārangi upoko: *contents*

"The inhabitants of this place invited us ashore with their usual Marks of Friendship, and shew'd us all over the place… and they had in it, Split and hanging up to dry, a prodidgious quantity of various sorts of small fish, a part of which they sold to us for such Trifles as we had about us…"

Captain James Cook, 1769

"…after having a little laugh at our seine, a common king's seine, shewd us one of theirs which was five fathoms deep. Its length we could only guess, as it was not stretched out, but it could not from its bulk be less than four or five hundred fathoms."

Joseph Banks, 1769

"…nor is the profusion more remarkable than the variety… shoals of the most excellent fish… our table was not only plentifully but luxuriously supplied…"

J.L. Nicholas, 1817

"In a few instants the crew had brought up on their lines an immense quantity of fish, which was exquisite eating…"

Dumont d'Urville, 1827

"…the fish itself, we found in considerable numbers strung along lines, dried, and smoked with the design evidently, of preserving them for a future day… eels are found in great plenty…"

William Barrett Marshall, 1836

"The fishing nets of the people are often of an enormous extent, and are generally made, by each family in a village… many of these seines, which are the common property of a village, are one thousand feet in extreme length…"

Joel Samuel Polack, 1838

"The natives catch large quantities of them [kahawai] with a bone hook at the end of a fish-shaped piece of wood, inlaid with the shell of the mutton-fish or haliotus, which bears the lively colours and brilliancy of mother-o'-pearl. This hook requires no bait, and a dozen of them are dragged along the water by a canoe which pulls at full speed through the shoal."

Edward Jerningham Wakefield, 1845

"The harbours abound in fish – abound is a poor word for it: they are literally alive with fish. M— and myself now live almost entirely on them at every meal; they are delicious, and in great variety. We have a fish here exactly like the salmon, and of as good flavour. On a sunny morning the surface of the harbour is a complete mass of fishy life…"

George Earp, 1853

"Every fish found in the surrounding sea is eaten by the natives...
The intimate knowledge the New Zealanders possess of the habits of
fish, and their success in fishing, are indirect proofs that much of the
ancient food of the people was derived from this source."

Arthur S. Thomson, 1859

"Very fine crayfish were taken in great numbers by diving, and sometimes by sinking baited wicker traps. Heaps of this fish, with mussels, cockles, and other bivalves, were collected in the summer, and prepared and dried for future use."

William Colenso, 1869

"...at Katikati harbour, one tribe of Natives
have a right to fish within the line of tide-rip;
another tribe of Natives have the right to fish
outside the tide-rip."

James Mackay, 1869

"I have seen them [blue cod] pulled up with lines, three or four to each, as rapidly as the baits could be fixed and let down. I believe four good fishermen could fill a whaleboat in three or four hours..."

W.H. Pearson, 1872

"...the minute detail with which natural objects have been discriminated and named... nearly every fish of large size or insignificant... that would remain unnoticed by most Europeans... have all distinctive Maori names..."

Sir James Hector, 1874

"They would go out into the deep sea, with their large canoes, for ten or more miles from the shore. Cod, snapper, and other large fish, in great quantities, rewarded their toil. In their nets, they take numbers of mullet, dog-fish, mackerel, and other kinds that are found in shoals."

Rev James Buller, 1878

"The Natives frequently capture them [grey mullet] on still, moonlight nights by paddling their canoes close to the banks of the streams; the fish are startled by the beat of the paddle, and, leaping up, fall into the canoe."

R.A.A. Sherrin, 1886

"...one cannot help feeling regret that the Maori knowledge of fish in this colony has not been more carefully conserved..."

R.A.A. Sherrin, 1886

"Bivalves are easily obtained from the sandy shores and mud banks in estuaries… great interest was taken in preserving the best parts of the shell-fish beds, and occasionally a chief would tapu them to prevent their being exhausted by being overworked."

Augustus Hamilton, 1908

"The fish caught during the day are cooked in huge ovens, over 200ft. in length and about 4ft wide… about twenty or thirty thousand fish, he says, are cooked in an oven…"

Augustus Hamilton, 1908

"A particular season was devoted to eels… the larger eels swarmed into the sea in countless numbers… even now when this occurs it is not difficult to get a dray load of these fish in a short while… the whole coast was carefully fished…"

J.G. Wilson, 1914

"I have known Maoris catch two or three hundred [kahawai] each on the incoming tide…"

Tāmati Poata, 1919

"...of these the most famous to outside tribes was the kōura, which, though found in nearly all fresh-water streams, could nowhere be found in such quantities as at Rotorua... the kakahi had the greatest reputation locally."

Sir Peter Buck, 1921

"Captain Gilbert Mair... told us of a great seine, a veritable *taniwha* of a net, that was made by natives at Maketu in 1885 under the direction of the chief Te Pokiha. Some hundreds of persons were engaged in the task of manufacture, and it was, as usual, made in many sections. These numerous sections, when completed were assembled and joined together... the result was a huge seine 95 chains (2090 yards) in length, well over a mile!"

Elsdon Best, 1929

"The larger canoes would contain thirty men, more or less, and... proceeded to the more distant fishing grounds, those furthest from shore. When going to the outermost grounds canoes started during the night, hence many persons were taken to serve as paddlers..."

Elsdon Best, 1929

"...fish have become extremely important. ...marine and freshwater stocks, once taken more or less for granted, are now seen to be both valuable and vulnerable..."

L.J. Paul, 1986

Whakarāpopoto: *summary*

…He tini noa iho nga kai a te Māori o te taha moana, e kore e tae ate tuhi katoa…
"…So many types of food were gathered by the Māori from the sea/seaside, it's impossible to write them all down here…"

Ngāi Tahu scholar Teone Tāre Tikao, 1850-1927 (Beattie 1939: 26)

I te wā i a rātou mā muia ana ngā tai o ēnei moutere e te tini a Tangaroa. I kite ā-whatu, i rongo ā-kiri, i whāwhā ā-ringa, i te mutunga iho ka hua mai ko ngā tikanga me ngā hangarau i āta whakawhanake hei mātauranga tuku iho ki tēnā, ki tēnā whakatipuranga. Hei aha rā? Hei whāngai i te tangata, ko tōna whānau, ko tōna hapū, ko ōna manuhiri. He mea hao te ika ki rō kupenga (kotahi maero te roa o ētahi), ki rō pouraka hoki. I werohia te ika ki te pātia, i hīia ake mā te poapoa me te matau. I waihangatia ngā taputapu hī ika ki ngā rawa māori pēnei i te rākau, i te toka, i te kōiwi, i te parāoa, i te anga hoki. I whiria ki ngā weu o te momo otaota pēnei i te harakeke me te tī. I āta waihangatia, i whakawhanakehia ēnei taputapu i ngā rautau maha.

Ehara i te mahi ngāwari te waihanga i ngā mata koi me ngā kāniwha hei wero, hei pupuri i te ika ki te matau ki ngā rawa māori. He mōwhaki nō ētahi o ngā rawa māori i pēnei ai. E rua ngā momo matau i hangaia e te Māori: i hīia te ika nō te wai hōhonu ki te matau; i hīia te ika nō waenganui i te wai hōhonu me te wai pāpaku ki te pā kahawai, ki te pōhau mangā rānei. He porowhita te hanga o te matau a te Māori, e anga whakaroto ana te mata, e mau hāngai ana ki te kauawhi, engari kei te koki hāngai atu i te aho. Ko te mea mīharo o tēnei tū matau, ka ngaua e te ika, ka whakamaua ki te kauwae o te ika, e kore e tuku. Kāore he take o te kāniwha. Me taikaha, me mātotoru te hanga o te kauawhi koi whati, koi riro tana ika. Me uaua te here i te mōunu ki te matau, i herea kētia ki te tuaina e mau ana ki te pūtake o te matau.

I whakamahia ngā konu me ngā rawa waihanga a te Pākehā e te kaihī Māori i ngā rautau 19 me te 20. Nā rā, ka ngaro ētahi o ngā taputapu i hangaia ki ngā rawa o nehe, ka ngaro ētahi o ngā tikanga hoki. Nō te hāwhe tuarua o te rautau 20 i haoa ngā wai o Aotearoa e te hunga hao ika arumoni nei. Nā runga i tēnā, i heke nui ai ngā ika, ōna rahi me ngā momo i ngā wai o konei. Nā te nui o ngā ika i ngā wā o nehe, i āhei ai te Māori te hī ake i ngā momo ika i tēnā, i tēnā tauranga ika mā te pouraka, mā te kupenga, mā te pātia me te matau. Nā te huringa kē o te noho ki Aotearoa nei me ngā ka ii puta i te mahi ahuwhenua a te Pākehā i te rautau 19, i heke ai te hī ika. Engari ko ngā whānau me ngā hapū kei waho i ngā tāone i mau tonu ai ki ō rātou tikanga hī ika. He nui ngā pārongo mō te ngaro haere o ngā momo ika, hāunga anō te rautau 19, engari kei te maumahara tonutia, kei te pupuri tonutia i roto i ngā kōrero ā-waha, ngā kōrero tuku iho ki Te Rōpū Whakamana i te Tiriti o Waitangi.

Ehara te ao Māori tūturu i ao tū noa. E kore e taea te kī nō te whakarere atu i ngā taputapu tahito, nō te rironga o te reo hī ika ki te pō, nō te mau ā-ringa i ngā taputapu me ngā hangarau a te Pākehā, kua ngaro te 'hī ika Māori nei'. I tautokona ngā tikanga whai hua a te Māori e ngā ariā me ngā hangarau a te Pākehā.

He rerekē ngā taputapu hī ika i tēnā rohe, i tēnā papakāinga. Me kī, kei ngā āhuatanga o te wā, o te moana, o taua rohe, o te tangata hī ika te tikanga. Ahakoa te rerekētanga o ngā rawa, o ngā taputapu, o te wā, arā kē ngā wāhanga o te matau e kore e rerekē. Nā reira i āhei tātou te āta titiro whakamuri ki ērā wā. He rerekē ngā matau mai i te wā tuauri tae atu ki te wā o te puāwaitanga. He rite tonu te hanga o ngā matau tūāuki ki ngā matau nō ngā moutere o Te Moananui-a-Kiwa. Ka huri te wā, i whakawhitia ngā matau tuauki ki ngā matau i whakarākeitia, i tāraia ki ngā kāniwha. I ahu mai tēnei tūāhuatanga i te tai whakararo o Te Ika-a-Māui. Nā te kaha mina mai a ngā tūruhi Pākehā i te mutunga o te rautau 19, he nui ngā kape matau i hangaia, me uaua te kite ko tēwhea te kape, ko tēwhea te matau tūturu. Nā te hei matau, nā te ngaro o ngā wāhanga rākau, o te muka, me te taenga mai o te matau rino, he mea uaua anō te whakamahuki i ngā kōrero mō te hangarau o te matau Māori i te katinga o te rautau 19 me te tīmatanga o te rautau 20.

I ngā rangahau mō te painga o ngā momo matau nō nā tata nei, i kitea he pai ake ko ngā matau hanga J rino nei i ngā matau hanga porowhita mō te hī i te tino maha o ngā ika. Engari, he pai ake te hanga porohita mō te pupuri i te ika. Ko te painga o te matau hanga porowhita ka pupuri i te ika e ora tonu ana, pērā i te hī aho roa, ā, koinei te take kua whakaara ake ngā kaihī o nāianei i tēnei momo hangarau nō tuauki. E hia kē mano tau te matau porowhita i whakamahia e ngā iwi taketake maha o te ao. Heoi anō, nō te keringa mai o ngā konganuku hei waihanga matau mētara, i ngaro ngā mātauranga e pā ana ki te matau porowhita. Ko tōna hua, kua hē ngā whakamārama a ngā tumu kōrero, a ngā kaiwāhi whenua mō ngā matau tuauki nei, kua kīia he matau kore take mō te hī ika, i whakamahi kētia mō te taki karakia. Koirā pea tētahi tūāhuatanga korokē, arā, he pōhēhē nō ngā kaihī o ēnei rā he tohu te matau porohita mētara o te hangarau o te ao hou anake. Engari kē, he whakahoutanga o tētahi hangarau tuauki.

Whāwhetai: *acknowledgements*

Focus for this publication is on the nature and function of traditional fish-hooks, the historical inter-relationships between Māori customary fishing, the development of commercial fisheries and their regulation by government, and the need to manage and conserve fish stocks in New Zealand following European settlement, as seen from my experience with research in fish taxonomy, New Zealand commercial fisheries, the Waitangi Fisheries Commission, and as a recreational fisherman.

The project was supported and encouraged by Carol Diebel and Dame Claudia Orange. Arapata Hakiwai, Rhonda Paku, Hokimate Harwood, Awhina Tamarapa, Dougal Austin, Huhana Smith, Matiu Baker, Ross O'Rourke, Ricardo Palma and Bruce Marshall assisted by providing access to collections, commenting on the manuscript, or engaging in discussions. Joan Costello and Paora Tibble provided translations of text in Te Reo and English. Harry Haase and Dr Gundolf Krüger (Göttingen University, Institute of Social and Cultural Anthropology), kindly provided images of hooks from the Forster collections.

Hemi Sundgren, Ron Lambert and Kelvin Day (Puke Ariki, New Plymouth), Michelle Horwood (Whanganui Regional Museum), Harry Allen (University of Auckland), Dimitri Anson (Otago Museum), Catherine Jehly and Chanel Clarke (Auckland War Memorial Museum), Larry Paul (NIWA), Keith Giles (Auckland Libraries), Steve O'Shea, Robin Watt, Roger Fyfe, and Janet Davidson, all provided access to collections of taonga, comments, or useful discussions.

A 2009 Winston Churchill Memorial Trust Fellowship and assistance from the New Zealand Lottery Grants Board enabled me to travel to Europe to examine fish-hooks collected by James Cook and other early explorers. Curators, keepers and managers of the museum collections, and their assistants, gave generously of their time in permitting access to collections and in discussing aspects of traditional fishing technology. They include Arina Lebedeva (St Petersburg, Russia), Peter vang Petersen (Copenhagen, Denmark), Phillipe Peltier (Paris, France), Gabriele Weiss

(Vienna, Austria), Mary Cahill, Fiona Reilly and Margaret Lannin (Dublin, Ireland), Sally-Anne Coupar (Glasgow, Scotland), Chantal Knowles and Brenda McGoff (Edinburgh, Scotland), Rachel Hand (Cambridge, England), Sian Mundell and Jeremy Cootes (Oxford, England), and Natasha McKinney (London, England).

Many thanks to the international institutions that supported this work by permitting me to photograph items from their collections (see p224), and by freely providing access to their digital images. My appreciation to Matt Lind, Peter McMillan and Paddy Ryan who also kindly allowed me to reproduce their photographs, and to Tim Chamberlain and Mandy Bayliss at Streamline Creative, who helped turn my manuscript and concepts for this project into the book you are now holding.

This publication originated from earlier collaborative studies with Dr George Habib (Tuwharetoa/Arawa) in the late 1980s. Dr Habib had been commissioned to prepare reports for the Waitangi Tribunal on marine fisheries used by Māori. He attended Tribunal hearings, commenting on the reliability of the evidence being presented, provided alternative interpretations where necessary, and overview summaries of the evidence presented to the Tribunal. Following his return to New Zealand in 2010 after working as an international fisheries consultant with traditional and developing commercial fisheries throughout the Indo-Pacific and Africa (including Kiribati, New Guinea, Kenya, Nigeria, Ghana and Sri Lanka), George expressed interest in working on this publication with me and in helping to make the information available to a wider audience. Tragically, he passed away in 2013 before he could begin work on reviewing the published documents and draft manuscript. *Haere, Haere, Haere.*

Funding for this book has been generously provided by the George Mason Charitable Trust and the Charles Fleming Publishing Fund.

Finally, special thanks to my wife, Kim Lund, for her support and encouragement throughout the project.

Chris Paulin

Kupu whakataki: *introduction*

He kai tangata, he kai titongitongi kaki;
He kai na tona ringa, tino kai tino makona noa!
"Food from another is little and stinging to the throat;
food of a man's own getting, is plentiful and sweet, and satisfying."

Māori have always fished: as the first inhabitants of Aotearoa (New Zealand), their survival depended on knowledge of the oceans and waterways that provided them with a readily available supply of protein and raw materials to support their daily life. Coastal fish stocks gave Māori a healthy staple diet and provided materials for building and clothing.

Māori fishing traditions and their close relationship with the sea can be traced back to the demi-god Māui who, according to legend, fished up Te Ika a Māui (the fish of Māui) – the North Island – using the jawbone of a dead ancestor, from which he created a magical fish-hook. Legend has it that Māui and his brothers were angling using Te Waka a Māui (the canoe of Māui), an alternative Māori name for Te Waipounamu, the South Island.

As early Māori had no written tradition, history and culture were passed from generation to generation through proverbs, song and legend and by other means such as whakairo (carving) and decorative tukutuku panels. This rudimentary and organic transfer of information between generations was critical to fishing practice.

To some Europeans, Māori fishing appeared unregulated and haphazard in nature; in fact there was a complex and often extensive set of rules and practices ensuring that fish were able to be taken sustainably in times of plenty and conserved when fisheries became depleted or during sensitive times such as spawning seasons. The use of rahui and tapu are examples of Māori cultural customs that were in widespread use and remain as important controls today.

The relatively small Māori population and the limitations enforced on Māori fishers by the lack of physical resources such as modern fishing equipment allowed them to enjoy an abundant source of protein without the problems of over-utilisation and exploitation. The rules surrounding the gathering of fish and shellfish were inherited and imposed by respected elders.

Māori were manufacturers and users of two principal types of matau or traditional fish-hooks. Bottom-dwelling fish were caught with the traditional suspended circle bait hook, and predatory pelagic or midwater fishes were caught with pā kahawai and pohau mangā, unbaited trolled lures. The circle-hook matau was not 'set' by the angler using a rod; instead, woven handlines were used to haul in fish that became self-hooked on the barbless inturned point.

The suspended matau was dismissed by European observers throughout the 19th and 20th centuries as inefficient and impracticable. However, the design has since been shown to be a masterpiece of form and function, more efficient than many modern hook designs. The rotating circle-hook design has been rediscovered and is now considered an innovation for improving landing rates in commercial pelagic longline and deepwater fisheries. This is because fish are rarely harmed by being gut-hooked and survival rates for returned undersized or non-target species can be improved.

French sailors assisted by Māori hauling fishing lines in from the sea at Kororāreka (Russell). Watercolour by Lauvergne, Barthelemy, 1805-1871. Alexander Turnbull Library, Wellington, New Zealand, Ref: B-098-005

A lesser known but also innovative practice saw large reef fishes targeted with small internal-barb hooks that were swallowed by the fish and caught on the branchial gill arches, using a unique hook design and fishing method. While this is no longer employed, it represents a previously unrecognised technological achievement.

Collections of matau are held by museums and private collectors around the world, providing a wealth of accessible historical information. In 1997 the first Māori Speaker of the New Zealand House of Representatives, Sir Peter Tapsell, observed that "each taonga's ancestral pathway has woven a pattern of human interconnections upon the land for generations, forming a korowai, or cloak, of knowledge". By exploring the evolution of matau, we can use these korowai to interpret traditional and modern fishing practices and gain a better understanding of the current state of fishing affairs as it affects Māori and Pākehā alike.

The transition from traditional materials to European metals and synthetics makes inter-pretation of early Māori fishing and fishing methods problematic. Sharp points and barbs required for piercing and holding the fish on a hook, as with present day metal hooks, could not easily be manufactured from natural materials, and could not always be relied upon, because of their often brittle composition. These were inclined to collapse under load.

There is extensive evidence of fish consumption in coastal middens and records of Māori fishing and fish names in early literature and in oral histories submitted to the Waitangi Tribunal. However, there is no comprehensive or detailed written record of Māori fishing activities pre- or post-European settlement. Regardless, there is enough evidence for us to conclude that Māori had a profound knowledge of the fish species available in coastal and inshore waters of Aotearoa.

From material in collections and historical records, a depth limit for Māori fishing of between 50 and 100m can be inferred but we cannot be certain of what Māori knew about fish from greater depths. We can conclude that they probably had little need to target deepwater fish species such as hoki, ling and groper, as they could readily catch sufficient quantities of desirable eating fish in much shallower water. However, these species should not be excluded from the diet of Māori as the depth range and distributions of these prized species has undoubtedly changed due to recent commercial and recreational overfishing. There are many records of these species being taken in near-shore coastal waters as recently as the 1950s and 60s.

Large oceanic species such as tuna, marlin or swordfish could contribute significantly to daily food resources but open-sea expeditions (out of sight of land), and even offshore fishing (beyond 5 kilometres) could be undertaken only in ideal weather conditions. The risks involved versus the benefits of such expeditions would seem to make them unnecessary and unlikely, given the access that Māori had to plentiful inshore fisheries.

This is not to say that Māori lacked the skills and technology required to catch and process large fish or large fish catches, as history shows that they clearly took enormous inshore catches of eel and mullet and other species when these were seasonally available. In fact the first 'fish factory' in New Zealand was undoubtedly operated by Māori working co-operatively, with coastal fish in an artisan fishery that survived and prospered until the arrival of European factory trawlers.

Ka pu te ruha ka Hao te rangitah!
"When the old net is cast aside, the new net goes fishing!"

Once Europeans became settled in New Zealand, Māori were quick to embrace the settlers' agricultural practices and more urban lifestyle. In some ways this enhanced their own living standards but it changed their culture irreversibly. Intensive agriculture and farming provided for stable production but Māori began to leave the countryside in search of work in an increasingly urbanised New Zealand.

This urban drift was offset for a time by fishing as it remained a core component of Māori life, but once European New Zealand was established, government and big business turned its attention from land acquisition and development and broad-scale agriculture to fishing. They came to realise that there was enormous potential value in the almost untouched New Zealand fishery that Māori had been quietly harvesting for centuries. Following the pattern of European land acquisition, successive New Zealand Governments actively sought to remove traditional Māori fishing rights using various methods including legislation, lobbying, misinformation, untruths and skulduggery.

As this went on, fishing tackle, once made of wood, stone, bone, ivory or shell, lashed with fibres from plants such as flax or cabbage tree, was steadily being replaced with equipment made from metals and synthetic materials. Wooden waka were being replaced by modern fishing vessels. Fish were and still are taken by line with suspended hooks and trolled lures, as well as nets, traps and spears but the huge hand-woven flax nets, some up to a mile or more in length, once used by local communities, had been replaced by synthetic monofilament nets operated by commercial trawlers and [some Māori-owned] fishing companies.

Traditional hooks were largely discarded and by the early to mid-1800s, much mātauranga or fishing knowledge had been lost forever. Some Māori continued to make fish-hooks and lures following traditional designs; however, the overwhelming number of mass-produced metal hooks imported by Europeans, combined with the difficulties of manufacturing traditional hooks, led to a gradual decline in production. A small industry emerged in the late 19th century satisfying demand for indigenous artefacts and this led to production of numerous replica and fake hooks, by both Māori and Europeans. These were sold to unsuspecting tourists and collectors as genuine Māori artefacts but this entrepreneurial enterprise did little to sustain the artisanal skills required for making effective fishing tackle.

Economic development and bicultural New Zealand

He kai kei aku ringa
"There is food at the end of my hands."

As with land development and agriculture, development of large-scale deregulated commercial fishing in New Zealand waters in the 19th and 20th centuries had consequences for Māori. Undoubtedly and somewhat unavoidably, the increase in fishing activity resulted in significant declines in fish numbers, distribution ranges and fish sizes. Expansion of European interests into commercial fishing in the late 1800s also led to increasing government regulation and conflict with Māori fishing rights.

These rights had been guaranteed under the 1840 Treaty of Waitangi and initially government regulations recognised and protected Māori fishing rights; however, competition for declining resources, initially concerning oysters and mullet in northern waters, soon led to claims by Europeans that stocks were being depleted by Māori.

In an ironic and short-sighted twist, the Crown banned under legislation techniques that Māori had used sustainably since their arrival in Aotearoa. Māori were required to obtain a licence when fishing for any reason other than for personal or family consumption. In early fisheries legislation, Māori rights were acknowledged by providing for coastal areas that were to be set aside for their fishing upon application to the Crown, yet no such areas were ever established. To make matters worse, although Māori continued to protest throughout the 20th century, customary fishing cases were treated as criminal prosecutions in the courts.

Rural Māori continued their fishing traditions and a wealth of information documenting the decline of fish stocks, not only in the 19th century but within living memory, is available in Māori Land Court documents and in the oral histories presented to the Waitangi Tribunal, an agency established to address Māori grievances and breaches of Māori rights under the Treaty of Waitangi. This information though was largely ignored by fisheries managers and Māori were significantly disadvantaged financially and spiritually.

This was to change somewhat with the recognition of the Treaty of Waitangi as part of New Zealand law in 1975, and with the establishment of the Waitangi Tribunal Fisheries Commission. This controversial legislative change led to some multi-million dollar settlements of Māori land and resource grievances and to a change in the status of Māori as part of New Zealand's cultural heritage. It brought slow – and in some quarters begrudging – recognition of the self-serving behaviour of the Crown historically toward Māori.

In spite of this fresh perspective in 1983, the New Zealand Government developed and implemented new wide-ranging fisheries policies and laws, failing to consult Māori. The

School sharks (*Galeorhinus galeus*) hung up to dry at Takahiwai, Whangarei Harbour, 1914. Sir George Grey Special Collections, Auckland Libraries, 7-A14507

government also staged a campaign to eliminate small-scale fishing operations from the fishing industry – Māori and non-Māori alike. The writing was on the wall for many fishing operations and large-scale trawling using factory vessels became the way forward for fishing in Aotearoa.

From the early days of European settlement, commercial harvesting rapidly exceeded the productivity of many fisheries. The once thriving commercial mullet fishing in the northern harbours collapsed before the end of the 19th century. Although some fisheries managers were expressing concern, it was considered by the Crown that overfishing was restricted to areas close to fishing ports. Government supported and encouraged major investment in the industry throughout the early 20th century but failed to legislate or implement any controls on the quantity of fish caught, other than through licensing of fishermen and vessels.

Deregulation in the 1960s, followed by the declaration of a 200-mile Exclusive Economic Zone in 1977, led to a rapid expansion and dramatic increase in commercial fishing operations both inshore and in deep water. By the 1980s, concerns surrounding collapsing fish stocks, overfishing and over-investment in the industry resulted in the introduction of the Quota Management System (QMS) in 1986. This was an attempt regulate and control the commercial catch. In the process, many part-time fishers, who were predominantly Māori, were excluded from the industry.

Kaua e mate wheke mate ururoa
"Don't die like an octopus, die like a hammerhead shark."

Persistent Māori claims and the establishment of the Waitangi Tribunal Fisheries Commission resulted in the introduction of the Māori Fisheries Act 1989 and Deed of Settlement in 1992, under which Māori agreed that all current and future claims in respect of commercial fishing rights were fully satisfied and discharged. This resulted in enormous economic benefits for some Māori, but the government of the day could not find a way to include disenfranchised urban Māori in this windfall. These Māori who had lost traditional tribal connections, ironically through urban drift following European settlement, felt slighted that government appeared to negotiate only with a few influential Māori and left allocation of fisheries to the Commission. Although it could be said that this undermined Māori collective activity and did not resolve the roles and inter-relationships between traditional iwi and urban Māori, it did open debate and revealed a number of rifts within the Māori community.

Recognition of Treaty rights through the Waitangi Tribunal has enabled Māori to attain greater involvement in New Zealand's fisheries as the industry has expanded significantly over the past few decades. Māori interests moved from minimal commercial involvement in the 1970s to now controlling over 40 percent of the industry. Customary fishing rights acknowledged and provided for in the Māori Fisheries Act 2004, enable development of systems and processes for government and Māori to work together to safeguard New Zealand's indigenous resources at international levels. The tension between the Crown's need to govern New Zealand's resources in the interests of all, and the right of Māori to exercise rangatiratanga (self-governance) is an ongoing sovereignty issue.

The relationship between Māori and Europeans and their shared use of the resources is dynamic and changeable. This is reflected in the difficulties of interpreting Māori fishing activities from various indirect sources of information, be it from oral traditions, archaeological evidence, museum artefacts, or historical and archival accounts that may reflect regional, localised and chronological variation. The extent to which Māori fishing activity changed as a result of European influence is subject to interpretation and debate.

Māori culture is not static; Māori have always been and continue to be a vital part of New Zealand life. Anchored in the past but enriched by exposure to other cultures, Māori culture has evolved and strengthened, while the involvement that Māori have with fishing changed as traditional equipment was discarded and new tools and technology adopted.

The demise of customary materials and their replacement by European metals and fibres, and more recently synthetic materials, is a natural evolution and a benefit of European settlement in the creation of New Zealand Inc. Some modern Māori communities continue fishing traditions

South Taranaki eel weir photographed around 1930-40. Puke Ariki PHO2009-355

that are based on careful observations of generations of fishers and mātaurangi Māori, while embracing new methodologies, particularly in lower socio-economic areas such as the far north and the deep south. Inevitably, this has led to a regrettable loss of traditional skills and knowledge, a pattern that can be traced through other indigenous cultures exposed to modern technology.

Our lives bear little resemblance to those of even our very recent ancestors. Technology has advanced so rapidly that things once considered to be science fiction are now commonplace. It is interesting to wonder what our tupuna (ancestors) would make of modern fishing gear, acoustic fish-finders, vessels, and weather forecasting technology. Would they look on modern fishing gear as an insult to the knowledge and techniques handed down from their ancestors, or would they embrace these changes as a natural progression and use them in conjunction with their hard-won knowledge and experience? I would suggest the latter.

Mark Fenwick
Te Atiawa and Taranaki

FIG.1 **View overlooking Wainui Inlet, Golden Bay, 1843, with Māori drying fish (including stingrays and barracouta) on racks. Watercolour by an unknown artist. Alexander Turnbull Library, Wellington, New Zealand, C-030-019**

1

Rourou kōpaki: *the Māori food-basket*

"For this scarcity of animals upon the land, the sea, however, makes abundant recompense. Every creek and corner produces abundance of fish not only wholesome but at least as well tasted as our fish in Europe: the ship seldom anchored in or indeed passed over (in light winds) any place whose bottom was such as fish resort to in general but as many were caught with hook and line as the people could eat, especially southward, where when we lay at anchor the boats by fishing with hook and line very near the rocks could take any quantity of fish…"

Botanist and explorer Joseph Banks, Cook's 1st Voyage, 1770 (Beaglehole 1962: 6)

Fishing has always been of major significance to Māori communities in Aotearoa New Zealand (Fig.1). Using knowledge accumulated over countless generations, they caught fish with nets (kupenga), traps (hīnaki), spears (pātia), and lines (aho), using suspended hooks (matau) and lures (pā).[1] Māori made fish-hooks from wood, bone, stone or shell, lashed with muka or whitau fibre prepared from flax, demonstrating sophisticated craftsmanship and superior design that enabled them to surpass the fishing expertise of any other culture. Their fish-hooks were as efficient at catching fish as any modern steel hook.[2]

The importance of sea-transport, and presumably fishing, to Māori was evident from the first European contact in 1642 when the Dutch explorer Abel Janszoon Tasman reported a massing of some 33 canoes in what is now known as Golden Bay (Fig.2).[3] Following European settlement of New Zealand, metal hooks soon replaced traditional hooks as Europeans considered the Māori fish-hooks "ineffective" and "clumsy". As traditional fish-hooks were discarded, much indigenous knowledge (mātauranga) surrounding their design, function and use was lost. As indigenous knowledge was regarded as sacred it was not readily shared, and Māori often deliberately made up explanations to please the insatiable curiosity of Europeans,[4] so that today it is difficult to accurately determine the full story and extent of Māori fisheries.

FIG.2 **Golden Bay during Abel Tasman's visit in December 1642 where Māori attacked a rowboat and killed four of its crew.** Alexander Turnbull Library, Wellington, New Zealand, PUBL-0086-021

Food resources

That Māori subsistence use of the sea ensured a sustainable and low-risk catch is demonstrated in their rich tradition of ritual restrictions (rahui), enforced by supernatural penalties and jurisdictions.[1, 5-8] Māori legends and the historical record are complemented by archaeological interpretation, which has emphasised extractive opportunism, especially in southern waters.[9-12]

Māori legends, names of people and places, and oral history accounts denote an abundance of fish. Numerous myths and legends focus on fishing: for example, the legend of the demi-god Māui Pōtiki fishing up Te Ika a Māui (The North Island – the fish of Māui)[13, 14] and Kupe discovering New Zealand while in pursuit of a giant octopus (Te Wheke o Muturangi). Many other legends detail the origins of fish and surround fishing activities, including the inventions of the barbed hook, eel and crayfish pots, and fishing lines.[15, 16]

Historical documentation provided by early Europeans from the voyages of James Cook[17-22] and other explorers provide valuable insights into Māori fishing expertise, while archaeological studies have revealed specialised fishing camps established in all coastal regions. Early settlers, historians and Māori scholars, including Ernst Dieffenbach,[23] William Colenso,[24, 25] Elsdon Best,[1] Tāmati Poata,[8] Sir Peter Buck (Te Rangi Hīroa)[26] and many others, described the accessibility, high biomass and stability of the inshore Māori fishery,[5] particularly in northern waters.

Early Māori found a plentiful food supply in the giant moa and other birds, including extensive breeding colonies of seabirds scattered throughout the country and on offshore islands, as well as numerous seal colonies in coastal regions. The increasing Māori population eliminated the slow-breeding moa and decimated many of the seal colonies, while seabirds were only seasonal in availability.[9, 27-29]

Archaeological studies show that localised extermination of fur seals (kekeno, *Arctocephalus forsteri*) chronologically matched the extermination of moa in some regions.[30] Sites excavated at Purakanui and Long Beach, Otago, occupied around 1300 AD, and radiocarbon dates of midden material for Catlins sites in Southland show that fishing began to replace moa hunting and sealing as early as 1350 AD.[31, 32] This scarcity of large prey, especially after 1500 AD, caused a stabilisation[33] or even a reduction[9] in the size of the Māori population in the southern part of the South Island. New Zealand was devoid of native land mammals (with the exception of three species of small bats[34] that were not large or numerous enough to be of any food value), and hunting pressure gradually switched to smaller birds, supplemented by dogs (kurī, *Canis lupus familiaris*) and Polynesian rats (kiore, *Rattus exulans*) introduced by Māori, and cannibalism.[35-37]

Dogs were uncommon and restricted to a luxury food resource for persons of high rank. Polynesian rats, although recorded as part of the daily food,[37-39] were similarly limited to chiefs or visitors[25, 40] and were of little importance owing to their size.[41] Evidence for cannibalism does not suggest any significant contribution to diet[35, 36] and may have been reserved for ceremonial occasions.[37] Both James Cook and Joseph Banks reported evidence of cannibalism (and observed it on at least one occasion) on their first voyage of exploration in Aotearoa in 1769-70, and noted that probably only enemies slain in battle were eaten.

During the visit of the *Resolution* to Queen Charlotte Sound on Cook's second voyage, local Māori were observed cutting up the body of a young man they had killed in a raid. Following this observation of cannibalism, members of the crew encouraged the practice to demonstrate it first-hand to Cook.[42] This had serious consequences for Captain Furneaux and his crew when the *Adventure* (separated from the *Resolution* during a storm) arrived in Queen Charlotte Sound a few days after the *Resolution* had left. Māori related to the unfortunate victim attacked and killed a ten-man shore party, and were preparing the flesh in earth ovens when discovered by others from the *Resolution*.[42]

As terrestrial resources were depleted, marine animals became the primary food resource and numerous fish species and other marine animals such as crayfish and shellfish provided the principal food supplies.[9, 12, 43]

Although fish and shellfish resources were abundant, the availability of carbohydrates from plants was more limited, and this may have restricted population growth in all areas except in northern regions where kūmara (sweet potato, *Ipomoea batatas*) could be grown.[44] Access to fat

FIG.3 Māori fishing settlement, possibly at Haumuri, North Canterbury. Watercolour by Frederick Weld, 1850. Alexander Turnbull Library, Wellington, New Zealand, A-269-012

or carbohydrate to supplement a protein diet is essential to avoid starvation, and in southern areas where kūmara could not be grown, successful permanent habitation depended upon a reliable source of fat that was available from sea mammals.[44] The abundant fish stocks, supplemented by kūmara in northern regions, and bracken fern-root (rarauhe, *Pteridium esculentum*) elsewhere, were sufficient to provide adequate food supplies for Māori, except when seasonal periods of adverse weather prevented harvesting activity, or changing climatic conditions influenced the abundance of some marine species.[44-47]

Habitat destruction and the vulnerability of New Zealand bird fauna to predation by both Māori and the introduced Polynesian rat soon led to depletion of many birds, with approximately 30 species becoming extinct between the time of the arrival of Māori and the arrival of European settlers in the 1800s.[48-50] Once these terrestrial species were overexploited, Māori were dependant on marine resources such as fish and shellfish. Communities became concentrated in coastal regions, with extensions inland[51] where rivers, lakes, marshes and swamps were full of small fish, including kokopu (*Galaxias* spp.), lamprey (korokoro, *Geotria australis*) and other freshwater species such as large rich eels (tuna, *Anguilla* spp.) as well as mussels (kākahi, *Echyridella* spp.) and small crayfish (kōura, *Paranephrops* spp.).[24, 52]

Māori fishing

Fishing, especially in the sea, was a common activity for Māori throughout the country prior to European settlement (Fig.3).[1, 8, 35, 53] The frequency with which shell middens, generally rich in fish remains, occur along the New Zealand coasts suggests heavy coastal exploitation, given low population densities of Māori.[10] Isotope studies have shown that about 90 percent of all food energy consumed by Chatham Island Māori was of marine origin (including fish, shellfish, and sea mammals[54]). Food of marine origin varied from 11 percent at an inland site at Rotoiti, eastern North Island, to 61 percent at a coastal site at Wairau Bar, Marlborough.[44, 55]

For Māori, the sea provided kai moana – literally 'food from the sea'. Marine habitats became the main source of protein and fat. Methods of procuring and preserving fish were developed over generations,[51, 56, 57] and the unwritten Māori lunar calendar marked the seasons of appropriate food supplies and which fish species were targeted at a given time.[38, 40, 51, 58]

In the early 20th century, Māori scholar Tāmati Poata described seasonal fishing in the Bay of Plenty region: February and March were the months for catching blue maomao (*Scorpis violaceus*) when they were so fat they were hard to grill, and kahawai (*Arripis trutta*), which were steamed in hāngī (earth ovens) and dried for winter food. March, April and extending into May were the months for fishing for groper (hāpuka, *Polyprion oxygeneios*); while snapper (tāmure, *Pagrus auratus*) fishing began in April – those caught earlier than March were not particularly cared for. Blue warehou (*Seriolella brama*) and moki (*Latridopsis ciliaris*) were taken in June and July; while August, September and October was the season for tarakihi (*Nemadactylus macropterus*), ngature (red moki, *Goniistius spectabilis*?), porae (*Girella tricuspidata*), blue cod (rāwaru, *Parapercis colias*), marblefish (kehe, *Aplodactylus arctidens*) and gurnard (kumukumu, *Cheilodactylus kumu*).[8]

Fish were prepared and eaten using a variety of methods. They were occasionally eaten raw or prepared as a surimi-like paste,[1] or steamed in earth ovens,[1, 56] grilled over open fires,[37, 56, 59] or wrapped in leaves and placed on hot stones.[60] Fish were often baked whole without removing the entrails.[18] The entrails of some species – such as marblefish – were particularly popular.[37] Sir Peter Buck (Te Rangi Hīroa, anthropologist, and later Director of the Bernice P. Bishop Museum in Hawai'i) described how in the right season, the entrails of marblefish were more esteemed than the flesh and documented the following proverb or whakatuki:

Hoatu ki te kainga	Go on to my home
Kotaku ika ki a koe	My fish will be for you
Ko te ngakau ki au	And the entrails for me[37, 44]

During periods when fish were abundant, the surplus catch was prepared and stored for later use. The most common practice was to clean and split the fish, which were then hung on high wooden racks or stages to dry in the sun. Alternatively the fish could be filleted, cooked, and then dried before being packed into kelp bags. Fish could also be placed in running freshwater for four days before being cleaned and dried. Fish prepared by these techniques could be stored for winter use, and would last for up to a year.[1, 56, 61]

Both William Colenso (a Church missionary who became a noted traveller and botanist) and Sir Peter Buck described an unusual practice of marinating crayfish (kōura, *Jasus* spp.) in running freshwater for about three to five days, after which the flesh would easily separate from the shell, keeping the whole body of the crayfish intact, including the flesh from the legs and antennae. The bodies were then used to prepare a pungent and popular food (kōura mara) which could be eaten raw, baked in an earth oven (hāngī), or dried on racks in the sun.[1, 25, 37]

Historical accounts

Following European contact, Māori fishermen throughout New Zealand seized the opportunity to trade; they bartered fish to explorers from the late 1700s, and to the developing European settlements in the 1800s.

On his first voyage of exploration into the Pacific in 1769, James Cook's mission was to construct an astronomical observatory in Tahiti to observe the Transit of Venus for England's Royal Academy (the expedition's sponsor), before travelling south in search of the Great Southern Continent, which was presumed to exist. While in Tahiti, Cook established good relationships with many of the islanders, and decided to take a young man named Tupaea with him as an interpreter and a guide because of his familiarity with the islands in the region. This proved fortuitous when Cook arrived in New Zealand, as Tupaea's knowledge of the Tahitian and English languages enabled Cook to communicate and trade with local Māori.

Although meetings with Māori were not always friendly, with Tupaea acting as an interpreter Cook was able to avoid the fate of Abel Tasman who had sailed into Golden Bay in 1642. Tasman's brief and violent encounter with Māori appears to have resulted from mutual cultural misunderstanding when the Māori challenge was mistakenly interpreted as a welcome. Tasman named the place Murderers Bay and it was 127 years before Europeans returned to New Zealand with Cook's expedition.[62]

The journals of James Cook, Joseph Banks, Johann and Georg Forster, and William Anderson from Cook's three voyages of exploration between 1769-79 on the ships *Endeavour*, *Discovery*, *Adventure*, and *Resolution* provide numerous references to fish and fishing activities in New Zealand waters.[17-21, 63]

Fish supplies were purchased from Māori at every opportunity.[17, 63-67] While in Mercury Bay, Coromandel Peninsula on 8 November 1769, Cook reported that for small pieces of cloth, the expedition obtained enough fish to feed all 94 men on board the ship, and that soon they had

FIG.4 **James Cook's astronomical tents in Ship's Cove, Queen Charlotte Sound, with Māori fishing parties.** Watercolour by John Webber 1809. Alexander Turnbull Library, Wellington, New Zealand, B-098-015

more fish than they knew what to do with. Later the same month, on anchoring near Whangarei, Cook named the location Bream Bay, noting that: "…we had no sooner come to anchor than we caught between 90 and a hundred breams…" [snapper]. On other occasions Cook ordered that no bartering with Māori (who were keen to obtain metal tools) could take place until they had brought sufficient supplies of fish to the ship.[17]

In January 1770, Cook arrived in Queen Charlotte Sound, Marlborough, to careen and clean the hull of the *Endeavour*, and replenish supplies in the calm and sheltered waters (Fig.4). While there, Joseph Banks spent time ashore collecting specimens of both plants and animals. He observed that the vegetation and number of terrestrial animals was a little disappointing, but noted: "…for this scarcity of animals on the land the sea however makes abundant recompense…" and made several observations on local Māori fishing activities.[19]

On 16 January 1770, James Cook himself described fishing with a small net in Ship's Cove: "…having the Saine [sic] with us we made a few hauls and caught 300 pounds weight [~136 kg] of different sorts of fish…"[17] These were identified by Johann and Georg Forster as representing 14 species. The abundance of fishes in the Marlborough Sounds was not uncommon for New

Zealand waters. The Forsters had earlier noted the fishing in Dusky Sound, Fiordland, describing hauling up fish as fast as the line could be let down, and referred to dining on fish every day while in New Zealand: "...we had nothing but fish in broath [sic] or chowder, fish boiled & fried, baked & roasted & in a pye [sic]: in short we lived like true ichthyophagi..."[21] Although the fishing was successful, by far the greatest quantity was obtained by barter with Māori.[18-20, 68]

In April 1772, the French explorer Marion du Fresne anchored his ships *Mascarin* and *Marquis de Castries* in Spirits Bay off the northern coast of the North Island. Jean Roux, an ensign aboard *Mascarin*, reported seeing several storehouses containing nets 80 to 100 fathoms (140-180m) long and 5 to 6 feet (l.5-2m) wide, weighted at the bottom by stones (makihea), and fitted at the top with light pieces of wood (korewa) for flotation.[66] Second in command Lieutenant Julien Crozet described how, on arriving at the Bay of Islands, the ships were met by Māori with gifts of fish: "...We had hardly anchored before a large number of canoes came off and brought us a quantity of fish, and which they explained they had caught expressly for us..."

Later the French traded nails for fish:

"...they brought large quantities of fish, for which we gave them glass trinkets and pieces of iron in exchange. In these early days they were content with old nails two or three inches long, but later on they became more particular and in exchange for their fish demanded nails four or five inches in length. Their object in asking for these nails was to make small wood chisels of them. As soon as they had obtained a piece of iron, they took it to one of the sailors and by signs engaged him to sharpen it on the millstone; they always took care to reserve some fish wherewith to pay the sailor for his trouble ..."[69]

After a month of friendly relations at the Bay of Islands, du Fresne offended Māori, possibly by fishing in an area where a rahui had been placed following a drowning of several members of the local tribe. On 12 June 1772, Māori warriors attacked, and du Fresne and 26 members of his crew were killed and eaten.[70]

In 1793, another French expedition led by Commander Antoine de Bruni d'Entrecasteaux on the vessels *Recherche* and *Espérance*, recorded Māori occupation of Three Kings Islands (Manawatawhi). D'Entrecasteaux speculated that such occupation was chosen because of the ease of catching fish. Later in the day he approached shore near North Cape and traded iron for fish-hooks, lines, sinkers and fish.[71]

Early European settlers were also impressed by Māori fishing equipment and methods, and often relied on them to provide supplies. In the first book to solely describe life in New Zealand, *Some Account of New Zealand, particularly the Bay of Islands, and surrounding country; With a description of the Religion and Government, Language, Arts, Manufacturers, Manners and Customs of the Natives, &c. &c.* (published in 1807), John Savage, a disgraced surgeon from New South Wales, noted that the area abounded in fish of all descriptions. He praised the excellence of the fish, crayfish, crabs and oysters and stated that the colonists would find fresh and saltwater fish in the "…greatest abundance and flavour…"

Savage spent two months at the Bay of Islands in 1806, and noted the expertise of Māori men and women in fishing – referring to Māori women as being "…as expert at all the useful arts of fishing as the men, sharing equally the fatigue and the danger with them upon all occasions…". He described the excellent quality and construction of Māori nets, lines, hooks and lures made using flax, and considered them better than those made of European materials available at the time. Savage expressed a desire to obtain some of the lines for catching tuna (Scombridae) and mahimahi (*Coryphaena hippurus*), large, pelagic (surface-dwelling) species with strong fighting qualities, on his return voyage to England.[60]

In 1814-15 John Nicholas reported seeing Māori at the Bay of Islands hauling ashore an immense net (kaharoa – a large drag net) containing snapper and other fish, which they readily agreed to exchange for a few nails. He noted that the people were "…very industrious in attending to their fisheries…" which were "…numerous and well supplied…" His narrative also records that Māori observed certain fishing rights with limits to areas marked by stakes driven into the water. Several rows of stakes were in evidence, delimiting areas belonging to the different tribes. Trespass on areas belonging to others was "…resented and met with vigorous opposition…" often resulting in inter-tribal fighting.[1, 59]

Dumont d'Urville, a French naval captain, visited New Zealand during three expeditions to circumnavigate the world and explore the southern oceans. On his second voyage, in 1827, he anchored his vessel the *Coquille* off Tamaki in Auckland and commented:

> "In a few instants the crew had brought up on their lines an immense quantity of fish… Every time we were becalmed the crew caught with lines an astonishing quantity of fine fish belonging to the species *Dorade unicolor*, [snapper] which are excellent eating. It is the same fish that Cook calls 'bream,' and appeared to be prodigiously abundant in these parts. Whilst we were at anchor off the Mogoia [Tamaki] River, the Natives of Tamaki loaded their canoes in the space of a few hours. To-day the crew soon caught hundreds, and there was enough to supply each plate with ample provision."[72]

During the third expedition of 1837-40, Charles Jacquinot, Captain of the *Zélée*, who accompanied d'Urville's *Coquille* (by then renamed the *Astrolabe*), visited Otago Harbour (Fig.5). Charles' brother, the zoologist Honoré Jacquinot – the ship's surgeon and naturalist – noted that:

FIG.5 **View of Otago Harbour during the visit of the *Astrolabe* and the *Zélée* under the command of Dumont d'Urville, 1837-40.** Alexander Turnbull Library, Wellington, New Zealand, PUBL-0028-181

"The beaches of the neighbourhood offer all the conditions required for seine-net fishing and fish is [sic] so plentiful that the net nearly always comes in full. The first time we used this method here, we caught such a lot that we had to throw several hundred-weight back into the sea…"[73]

European settlers described how the shellfish and crustacea found in New Zealand waters would compete with any found elsewhere.[24, 67, 74] Joel Samuel Polack arrived in New Zealand in 1831. He observed large Māori fishing parties, often involving several villages, and later described his experiences, commenting on the plenitude of the fisheries.[67]

"In ichthyology few coasts in the globe possess a greater abundance and variety than the shores, rivers, creeks, &c., of New Zealand; these are equal in taste and flavour with those of Europe. Large shoals of various species visit the coasts at certain seasons; at which period the Natives take advantage, and procure for themselves, with gigantic seines, a sufficiency of this favourite food for the winter season. Some deep banks lie off the east coast on which the *kanai*, or mullet, *wapuka*, or cod-fish, and the *kahawai*, or colourless salmon, abound. All visitors to these shores are unanimous in their praises of the flavour and variety of the finny tribes of this section of the Pacific, which are only found in salt water. The *patiki*, between the large flounder and the sole, is equally excellent with the European fish, as are also the *mackarel*, of which there are several varieties. Many other fish are equally numerous, answering to our *hakes, tench, bream, snapper, haddock, elephant-fish, pollock, salmon, gurnards, pipe-fish, parrot-fish, leather-jackets, cole-fish, John Dorys, sword-fish, cod*; various kinds of *skate* and *cat-fish, sting-ray* and *dog-fish*. Many of these in flavour and weight yield to none of their kind in any part of the world. Among the leviathans, who sport in shoals around these shores, the para paraua, or *spermwhale* (*Physeter macrocephalus*); the tohora, or *right-whale* (*Balsena mysticoetus*); the mungu nui, or *black physeter*; the *sun-fish* (*Diodon mola*); *fin-back* (*Balaena physalis*); the *musculus*, or *large-lipped whale*; frequent the coast in vast number. The mango, or *shark* (*Squalus*); *pilot-fish* (*Scomber ductor*); *flying-fish* (*Exocoetus volitans*); *hammerheaded shark* (*Squalus zygaena*); frequent the coast, especially the River Thames, in vast shoals, and are preserved by the Natives as winter food. The banks off Cape Brett produce shoals of cod-fish in the season, which also form a variety for the winter provision, together with a quantity of small flounders. In testaceous and crustaceous fish this country will also compete with any other. The beaches on the whole of line coast, however rocky, afford sufficient space for *clams, mussels, limpets, wilks* [sic], *cockles, sea-ears, sea-eggs, starfish, cuttles, crays*, and *oysters*. Some gigantic mussels grow to a foot in length: these latter are found in an upright position. Between the rocks are found a great variety of shell-fish, principally univalves, either turbinated or cockleated. The fish within them are of excellent quality... These banks often extend some miles, and are well filled with clams and cockles of great variety; and large oysters, but of very strong taste, from their beds being in the vicinity of the roots of the mangrove-tree, whose bitter leaves and seed afford some nourishment to the fish. Of course none but bivalves take up their quarters in the luxuriant mud, which is of the most fattening quality. Many varieties of the oyster cling to the rocks, whose taste is delicate. The pawa, or *mutton-fish*, also clings to the rocks. Among the crustaceous genus, a large kohuda, or *crayfish*, equal in flavour and size to our lobsters, is found in plenty among the rocks below the tide; the natives feel for this fish with their feet, and, with a sudden jerk, eject it from its quarters into a basket. The common *crab* exists in numbers, and the quantities of *shrimps* and their family are unbounded..."[67]

Other European explorers in the 19th century, including Ernst Dieffenbach[23], William Colenso[24, 25] and Thomas Brunner,[75] made observations on the abundance of fish and fishing activity and its importance to Māori. Missionaries including William Yate,[76] the Reverend James

Buller[77] and Samuel Marsden,[78] and settlers George Earp,[79] Edward Tregear[80, 81] and Edward Jerningham Wakefield[82, 83] all published accounts of their travels and provided anecdotal comments on the ready availability of fish species in various locations.

William Colenso described Māori fishing in about 1840:

"They were not (as many have rashly supposed) deficient in food... They were very great consumers of fish; those on the coast being true Ichthyophagi. The seas around their coasts swarmed with excellent fish and crayfish; the rocky and sandy shores abounded with good shellfish; the cliffs and islets yielded plenty of mutton-birds, and fat young shags and other seafowl, and their eggs, all choice eating. Sometimes they would go in large canoes to the deep sea-fishing, to some well known rock or shoal, 5 to 10 miles from the shore, and return with a quantity of large cod, snapper, and other prime fish; sometimes they would use very large drag nets, and enclose great numbers of grey mullet, dog-fish, mackarel, and other fish which swim in shoals; of which (especially of dog-fish and of mackarel), they dried immense quantities for winter use. They would also fish from rocks with hook and line, and scoop-nets; or, singly, in the summer, in small canoes manned by one man and kept constantly paddling, with a hook baited with mother-of-pearl shell, take plenty of kahawai; or with a chip of tawhai wood attached to a hook, as a bait, they took the barracouta in large quantities. Very fine crayfish were taken in great numbers by diving, and sometimes by sinking baited wicker traps. Heaps of this fish, with mussels, cockles, and other bivalves, were collected in the summer, and prepared and dried; and of eels also, and of several delicate fresh water fishes, large quantities were taken in the summer, and dried for future use..."[24]

In 1880 Charles Heaphy described making frequent fishing trips to haul fishing nets at Lowry Bay in Wellington Harbour, taking kahawai, moki and flounder. He described how great quantities of fish were dried in the sun on tall scaffolds near the pā at the mouth of the river.[84] Edward Tregear[81] described the huge nets made by Māori and frequently commented on the size and quantity of fish around the coasts, including groper of 50 lb (~20 kg) in weight.

Observations in the latter part of the 19th century record large fishing expeditions, often involving groups of over 1000 people. Richard Matthews recorded an account of a night of shark fishing at Rangaunu Harbour, Northland, in January 1855 involving over 50 canoes and some 1000 tribal members. This resulted in a catch of over 7000 sharks, including one large canoe that took six tons (6.1 tonnes) of smoothhound (kapetā, *Mustelus lenticulatus*) and bronze whaler (tōiki, *Cacharhinus brachyurus*).[87] In the same year Gilbert Mair reported the use of a huge seine net which: "...not less than a thousand persons were unable to haul..."[85, 86]

Other writers noted the use of huge seine nets of several thousand metres in length, with catches frequently measured in tonnes.[1, 67, 76, 85, 87-89] One anonymous account from about 1871 described hauling a net 86 chains long and 30 wide (1700 x 600m):

"As the net neared the shore a large number of men swam around the net on the seaward side to endeavour to prevent the escape of the kahawai by jumping over the top of the net, while a number of natives were on the landward side engaged in killing the sting rays and sharks of which some hundreds were taken besides araara, schnapper, taharangi, kumukuma and other fish... It was computed that 20,000 (twenty thousand) kahawai weighing about fifty tons was taken in this, the first haul. The kahawai were distributed among the tribes, large hangis (earth ovens) were made along the shore and when the hangis were ready the fish were packed in, and allowed to cook for 24 hours, after which the fish were placed on stays to dry by the heat of the sun. When perfectly dry they were stowed away, and in that state would keep for some years... The share allotted to me was about one and a half tons which were accepted and returned."[90]

Although these observers made numerous anecdotal comments on fishing activities (Fig.6), generally little attention was paid to the details of how Māori fishing gear was used and few noted the details of how fishing nets (kaharoa, kupenga), lines (aho) and hooks (matau) were made. The information was not recorded until Sir Peter Buck (1877-1951) and Dominion Museum ethnologist Elsdon Best (1856-1931) documented detailed observations on the methods of making and use of fishing nets, traps (hīnaki) and hooks in the early 20th century. This was at a time when the techniques were becoming obsolete as new European materials replaced the traditional harakeke (flax), bone, shell, stone and wood. As the design of the Māori fish-hook differed so greatly from the designs of metal hooks that Europeans were more familiar with, many hooks were described as "impossible looking" and historians classified them as amulets or luck charms.[101]

Decline of Māori fishing activities

The gradual decline of Māori fishing activity occurred during the mid to late 19th century when European food crops such as wheat, maize and potatoes, and domestic stock including chickens, goats, pigs, sheep and cattle became available for growing and rearing. The increasing availability of European crops and livestock reduced Māori dependence on fishing and a more agricultural lifestyle was adopted, especially in the northern regions, while in cooler areas further south fishing remained the primary food source.[37] New European laws in the latter half of the 19th century restricting Māori access to fisheries resulted in profound social and economic changes.[26, 91-94]

FIG.6 An amalgam of five or six sketches of incidents at Onehunga, including two men carrying a pole of fish over their shoulders, Māori in boats and carrying fish. Watercolour by Edward A. Williams, 1864. Alexander Turnbull Library, Wellington, New Zealand, B-045-001

In the second half of the 19th century, Māori continued with their traditional fishing practices, but communal village life was being steadily eroded by European influence. In 1865 the government made a deliberate attempt to individualise Māori property through the Native Lands Act, which sought to undermine the common ownership that was fundamental to Māori culture.[95] Around 1900 a significant change in the European perception of Māori was occurring. Māori village life, communities and economies were being absorbed into the towns and cities, a process of urbanisation that continues today. It was not until the late 19th and early 20th centuries, during a period of Social Darwinism,[25, 96-100] that historians made further observations on aspects of fishing or fishing equipment,[86, 101-103] and although attempts were made to collate the available information,[1, 37] much had already been lost.

Tribal fishing and smaller scale group fishing for family needs declined from about 1885, and large communal efforts had virtually ceased by 1910,[1, 91] although individual Māori fishermen still sustained family livelihoods from their personal efforts, and many coastal hapū maintained a fishing tradition as a significant part of their lifestyle and sustenance. The expansion of Māori farming and agriculture using introduced domestic stock and cultigens, and trade with Europeans for other resources such as kauri gum and flax, diverted attention from the sea, and as a result, fishing traditions, rituals and even associated language, were no longer as widely observed.

Sir Peter Buck noted the cultural changes, commenting that following the European fishing technique of "…careless leaving to chance, or indiscriminate dropping of a baited line in the hope of hooking anything that came along… rightly regarded by the Māori as the action of a kūware – a person devoid of practical sense, instead of implementing the Māori knowledge, traditions and lunar calendars passed down orally through generations, marked the degradation of barbaric culture and the advent of a higher civilization…"[37]

In the 20th century, traditional fishing knowledge amongst an increasingly urbanised Māori population became obsolete and almost disappeared due to lack of participation in fishing activity, and only snippets of indigenous information survived to be recorded and documented. Despite this, many coastal-dwelling hapū (in contrast to urbanised Māori), have retained their fishing traditions into the 21st century, with many customary materials replaced or complemented by European metals, and more recently, synthetic materials.[91-93, 104]

Documentation of traditional fishing

The mātauranga (knowledge and traditions) of fishing practices, passed down orally, have been documented by writers including William Colenso,[24, 25] William Travers,[39, 105] Johann Wohlers,[106] Walter Buller,[77] Alfred Grace,[97] Elsdon Best,[1, 15, 40, 107, 108] James Beattie,[109] Sir Peter Buck,[51] Henry Grey,[110] J.D. Peart,[111] Tāmati Poata,[8, 56] and Teone Tāre Tikao.[14] Cultural traditions, passed down orally through word-perfect rendering of karakia (incantation or chant) by ritual specialists or tohunga, have been subject to interpretation or distortion by European perspectives[26, 96, 112-116] during the process of documentation.[80, 81, 100, 105, 117]

Mātauranga Māori underpinned the tangata whenua's way of life. It includes language (te reo), indigenous environmental knowledge (tāonga tuku iho, mātauranga o te taiao), indigenous knowledge of cultural practices, such as healing and medicines (rongoā), cultivation (mahinga kai) and fishing (kai moana).

A disjunct between 'factual science' and mātauranga is often claimed where factual accuracy is only superficial. Indigenous knowledge is often treated with scepticism because traditional accounts indubitably make sense to indigenous people, in order to form a spiritually nurturing and holistic world view.[118, 119] Mātauranga encompasses not only language, cultural practices, ceremonies and indigenous knowledge, but all forms of art and cultural values.

Some interpretation is necessary as sayings (pepeha, whakatauki or karakia) were not always statements of fact and without clues to the figurative meaning, the true meaning can only be guessed.[58] William Colenso and Governor Sir George Grey were well acquainted with Māori folklore and proverbs, and they recorded the text of many karakia in the late 1800s. The attendant detail of how the karakia were used in the realities of daily life was not recorded,[24, 25, 120] hence much of their value for study, and the full meanings of the karakia used in everyday life, have been lost.[58, 121] Māori were cautious of the potential and the pitfalls of publishing their highly prized oral inheritance, but neither they, nor Europeans, could write manuscripts or publish karakia without alteration to the form, meaning and practices of the oral tradition.[122, 123]

The loss of mātauranga knowledge extended to fishing equipment, including hooks. Today the term *matau* is used generically for Māori fish-hooks,[124] despite the wide variety of shapes and materials used. As with other cultures throughout Polynesia and with present-day commercial and recreational fishers, Māori had a rich terminology surrounding fish-hooks and different hooks were used in specific habitats to target particular species.[125] Augustus Hamilton (1853-1913), Director of the Colonial Museum, and Elsdon Best recorded a number of additional Māori names for fish-hook including kawiti, maka, mākoi, matika, matikara, noni and reke,[1, 101] but today it is not known which variety of hook these terms may have referred to.

In recent decades there has been an increasing number of oral history projects recording the relationships between hapū and their fishing resources.[55, 91, 93, 104, 126, 127] These are also far from complete. For example, the importance of the fur seal has not been documented in Māori oral histories of the pre-European period[30] and fur seals were not mentioned in accounts of Māori hunting and fishing methods by Elsdon Best[1] or Sir Peter Buck.[37] Hunting methods and utilisation of fur seals by South Island Māori have been described in detail,[5, 128] and the taking of South Island fur seal pups through summer has been noted for seasonal harvesting of food resources,[129] but older animals were not mentioned. This record is inconsistent with the archaeological evidence that indicates both pups and adults were taken at breeding colonies.[28]

Mātauranga Māori and academic evidence are not necessarily in conflict: it has been argued that the regulatory and ritual emphasis of recorded Māori tradition are not incompatible with an ecological construction of the archaeological data, as a shift from periodically unpredictable resources to other high-yielding and more predictable resources may have had a religious explanation and context.[5]

Māori oral tradition is not well understood by European culture,[130-132] and some European interpretations and documentation of Māori tradition have been described as artificial historical interpretations which are neither accurate nor impartial.[133, 134] Ngāi Tahu kaumātua, Tipene O'Regan, in reference to the creation of the 'Great New Zealand Myth' and the associated stories of the arrival of Māori in New Zealand in a great fleet of seven canoes by Percy Smith and his colleagues in the early 20th century, as well as more recent interpretations of authentic traditions in the 1990s, stated:

"In the learned journals, in the formal publications of museums, in books standing on library shelves, we now have available for the general public, a considerable volume of what is, in effect, mystical and invented nonsense…[a result of]…cultural identity fed with inadequate and wrongly-based knowledge…"[95]

Stories passed down in non-literate societies as oral traditions to explain history and the observed natural world are subject to compression and stylisation. As a result, some oral histories[54] have been dismissed by scholars as unreliable:

"A modern anthropologist could possibly be forgiven for the relatively minor attention given to fishing behaviour… It is a subject field with certain difficulties – we are all familiar with the notion of the extravagant fish story in European society… Fishing is one of the most important domains of the apocryphal story and it would be wrong to think that twentieth century European fishermen have a monopoly on fishy folk tales…"[44]

Events may be transformed into stories that offer myth-like explanations and may undergo major changes to allow transmission through generations and, in some cases, to serve political or religious purposes.[135, 136] While oral tradition may be susceptible to manufacture, its worth is in the essential message it imparts and the power of the written word to entrench error makes criticism of oral tradition seem small.[112]

Customary Māori uses of the sea recognised ritual restrictions enforced by rules and penalties, whereas archaeological interpretations of Māori fishing behaviour emphasised extractive opportunism and foraging theory.[5] There are numerous accounts in the literature of Māori fishing activities, and extensive archaeological investigations and analysis, both of which largely express the Pākehā or European view, which often makes it difficult to determine the Māori perspective.[112] Evidence that cannot readily be understood should not be dismissed as a spiritual matter or as metaphysical; rather, a full interpretation and understanding of its meaning is required to reconcile the two approaches of mātauranga Māori and European scholarship.[2, 60]

2

Hīanga: *catching fish*

"It is a much-to-be-regretted fact that we know but little of the pursuit of fishing as practised by the Māori; its methods have never been explained by any of the early writers. The voluminous ritual and innumerable beliefs and superstitions pertaining to fish and fishing are practically unknown to us. A few fragments have been collected; the greater part of such lore has been lost. E taea hoki te aha? The salved fragments must suffice…"

Ethnographer Elsdon Best, 1924 (Best 1924b: 398)

The ancestors of New Zealand Māori originated from eastern Polynesia and arrived in Aotearoa between 1200 and 1300 AD[36, 137, 138] with a neolithic culture that lacked the technology required to refine and smelt metal ores. Consequently, Māori used natural materials including wood, bone, stone and shell to make hooks and other equipment to harvest fish, their main source of food. Unlike metals that can be forged and made malleable to be bent into any desired shape, or tempered to improve hardness, natural materials impose limitations through the nature of their structure. Hence any fish-hook manufactured using wood, bone, stone or shell had to be a compromise between the limitations of the material used and the desired hook shape (Fig.7).

New Zealand plants such as flax (harakeke, *Phormium* spp.), cabbage tree (tī, *Cordyline* spp.) and kiekie (*Freycinetia banksii*) provided fibrous material for fishing lines and nets. These fibres were described in the early 1800s as equal to or superior in quality to the jute, hemp and sisal then in use by Europeans.[17, 67] Joel Polack noted that "…Their fishing lines are infinitely stronger, and fitted to bear a heavier strain, than any made from European materials…"[67] Lines, fished with up to 10 hooks on each, were made from prepared flax (muka or whītau) by rolling the fibres on the bare thigh with the palm of the hand. Two lengths of rolled twine (takerekere) were then rolled together to produce two-ply twine (karure), and many different forms of twine and cordage could be produced with varying numbers of strands. Lines made from ti or cabbage tree were more durable than those made from flax, but were more difficult to make.[1]

In contrast, traditional matau are shaped in a manner which makes it very difficult for Europeans familiar with metal hooks to imagine they could ever be effective in catching a fish.[139] Although early European explorers, settlers and historians recognised the superiority of the fibres used for nets and lines, many suggested that Māori fish-hooks were odd, of doubtful efficacy, very clumsy affairs, or even impossible looking (Fig.8).[67, 96, 101, 140]

Prior to European contact, Māori food supplies were mainly fish and birds, supplemented by fern root and kūmara where it was available. In all regions, fishing was of primary importance and it is unlikely that Māori could have afforded to be dependant on any fishing technology that was in any way inefficient. Furthermore, the unusual circular design of the Māori hook was used throughout New Zealand. There must have been some distinct advantage in using these hooks in preference to other possible designs.[1, 2, 26, 35, 141]

Following the exploration of New Zealand by James Cook and other Europeans in the late 1700s, sealers and whalers began visiting the region and trading extensively with Māori for provisions and other services, providing metal tools (including large numbers of metal fish-hooks), as a form of currency.[1, 37, 42] The superiority of metal for working implements quickly became apparent, and stone, wooden or bone tools were rapidly discarded, in a process described by Sir Peter Buck as "…a feverish desire that spread like a pandemic…"[37]

Māori use of traditional materials to make fishing equipment rapidly declined in favour of other fibres,

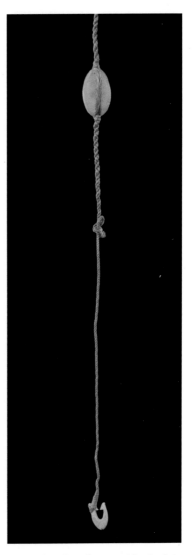

FIG.7 **Fishing line of prepared flax (muka) with bone fish-hook and quartz sinker. Possibly of Cook voyage origin. Hook 50 x 30 mm. Museum of New Zealand Te Papa Tongarewa, Wellington, ME012118**

metals and more recently, synthetic materials. However, they continued to replicate traditional forms in preference to the hook shapes introduced by Pākehā well into the late 1800s.[142] Māori were quick to adopt new technology as it complemented the existing fishing techniques, adding to, rather than replacing traditional fishing methods. Eventually the overwhelming number

FIG.8 **Composite fish-hook made from using a bone point lashed to a wooden shank. Possibly of Cook voyage origin. 128 x 71 mm. Museum of New Zealand Te Papa Tongarewa, Wellington, ME002496**

of mass-produced European hooks and the difficulty of making hooks from available metal diminished the use of traditional designs. *Ka pu te ruha ka Hao te rangitah!* When the old net is cast aside, the new net goes fishing!

European settlers began arriving in New Zealand in the early 1800s. They were primarily focused on farming and showed little interest in fishing activities. Consequently, details of how Māori fishing equipment was made and used were not widely documented. General observations of Māori fishing activities were recorded by explorers and early settlers,[23-25, 67, 76, 85, 89, 143] but by the end of the 19th century historians were beginning to note that much fishing knowledge had already been lost. It was not until the early 20th century that Elsdon Best prepared what is arguably one of the most important records of Māori life and culture.[1, 35] Best himself noted that little information on Māori fishing had been recorded and that there was not much he could do to supply the deficiency, stating that his work was "… a dashed bad paper – utterly inadequate – but I know nought about fishing…"[144] Although Best, and subsequently Sir Peter Buck,[37] gathered some useful information, particularly with regard to net-making, the European attitude is clearly demonstrated by Best's comment on night fishing:

"…the peculiar term *mangoingoi* was applied to fishing from the beach by night, but as to why anyone should so fish at night, and also claim a specific term for doing so at unholy hours is more than I can say…"[1]

As commercial and recreational fishermen know, some of the best fishing is at night. Night fishing, and other fishing activities that were rarely encountered and only reported by Europeans by chance,[86] were extremely important to Māori.

The prolific fish stocks that existed around New Zealand were described by early European explorers in the late 1700s, as well as the settlers in the 19th century. For example, Gilbert Mair, a sailor and merchant trader, remarked on a day's fishing in the Bay of Plenty in 1871:

"…in four or five fathoms water, with six lines, we had a whale-boat half full in an hour. The first fish hauled in were followed to the surface by swarms of snapper, kahawai, kingfish, barracoota, [sic] and maomao, and then we simply bobbed for them as you would for minnows in a brook until my arms ached with the exertion of lifting them over the boat's side…"[85]

In the late 1800s and early 1900s, newspapers frequently published fishing reports, which today would be considered apocryphal:

"…no other fish inhabiting our coasts can be captured in such quantity during the summer season as the groper… there have at times appeared immense shoals of the fish at or near the surface of the sea, so that a boat could not be rowed among them without striking them with the oars…"[1]

---

1 *Otago Witness*, 23 January 1890: 17

"There were seven schnapper [sic] lines on board, and we seemed to have stopped in the midst of a big shoal of schnapper and yellow-tail, with a few gurnet [sic] as policeman to keep the rest in order. The first catch was a four-pounder [1.8 kg] schnapper – and after that there was no cessation. More often than not two fish came up on the one line, and schnapper after schnapper flew into the big box we had. The coach driver was perched up on the deckhouse, and he rained down schnapper and yellow-tail on to the heads of those below, and on every side the fish came in, till the owner of the launch got tired of putting them in the box. In less than two hours over 2 cwt [100 kg] of fish had been caught…"[2]

These coastal fish stocks have been greatly reduced by modern fishing technology, particularly in the latter half of the 20th century. This has resulted in the biomass of certain species being reduced by up to 95 percent in some areas.[145, 146] As a result, the perception of the marine environment today is very different from what it was in pre-European contact times and in recent years many well documented reports of pre-European Māori fishing in the early literature have been improperly dismissed as extravagant fishy stories.[44]

Indigenous knowledge – mātauranga Māori
Mātauranga provided the scientific practice and technology required for Māori to survive in New Zealand before the arrival of European explorers and settlers with their still-developing Eurocentric scientific perspectives. Mātauranga Māori requires that species are valued for their practical benefits and not viewed simply as resources to be exploited, as their efficacy depended on their mauri or physical and spiritual well-being that is considered inherent in all things living or dead. Utilisation of those species required Māori to be responsible for that mauri, an obligation known as kaitiakitanga.[147]

Māori practices of food preservation which included dehydration (e.g. fish and shellfish on racks), exclusion of air (e.g. tītī – muttonbirds, *Puffinus* spp. and Polynesian rats – sealed in their own fat), and cool storage (e.g. rua kūmara, underground) or in pātaka. These practices were soundly based on scientific principles that food scientists now recognise.[148] Early accounts also describe the Māori method of preparing fish paste known as kaniwha (usually using snapper),[1] in a process identical to the present day preparation of surimi.[149]

In pre-European times, Māori lived mainly in coastal villages where harvesting the resources of the sea relied on understanding and overcoming variable currents, waves and tides, and required communal effort (Fig.9). The sea provided more than just the main source of food: even fuel in the form of driftwood was easier to gather from the beach than wood from the damp bush. The

2 *Otago Witness*, 18 December 1907: 19

FIG.9 **Māori men fishing for crayfish, taken possibly on the East Coast of the North Island, circa 1905 by Frederick Ashby Hargreaves. Alexander Turnbull Library, Wellington, New Zealand, 1/1-002601-G**

profusion of fish stocks in shallow coastal waters around New Zealand made it unnecessary for Māori to venture beyond the immediate coastline to meet their daily dietary requirements, even though they clearly did so at times.

Māori fishing was not based on chance or luck.[37, 103] They made their hooks to target particular species, and each species was fished in well-known habitats, often seasonally, based on the accumulated knowledge of generations of fishermen.[8, 56] If fishing was random, then a wide range of species would be expected to be represented in middens; however, across New Zealand only six species account for over 85 percent of the total numbers of identifiable fish remains reported in archaeological sites.[44] Of these, four are coastal reef-dwelling species: blue cod, snapper, spotty (paketi, *Notolabrus celidotus*), butterfish (marari, *Odax pullus*); one is a coastal soft bottom-dwelling species: red cod (hoka, *Pseudophycis bachus*); and the most numerous, a coastal/oceanic pelagic species: barracouta (mangā, *Thyrsites atun*). It should be noted that all six species have distinctive and readily identifiable skeletal elements, and the percentage of identifiable compared to unidentifiable fish remains that have been examined in middens is unclear.

Many studies suggest highly specialised and selective fisheries targeting a limited number of species. In the South Island for example, barracouta, red cod, wrasses, blue cod, ling (hokarari, *Genypterus blacodes*) and groper dominate middens.[92, 93, 150-152] The presence of less desirable species, such as inshore reef-dwelling labrids, including spotty, banded wrasse (tāngahangaha, *N. fucicola*) and scarlet wrasse (pūwaiwhakarua, *Pseudolabrus miles*), may result from poor sea conditions limiting access to preferable but less accessible fishing grounds at certain times.[153]

Isotope studies compared otoliths (ear bones) from red cod recovered from prehistoric middens at the mouth of the Shag River in North Otago with modern examples to determine the season in which the fish had been caught. The results showed that seasonal estimates from the prehistoric otoliths matched the seasonality of the modern fishery.[154] The fishery for groper was also seasonal, and the recorded description of this is closely tied to the known migratory pattern of the fish. Groper, or hāpuka, inhabit shallower inshore waters and outlying reefs from October to May, and in June to July they shift out to deeper water to spawn.[155] Traditional hāpuka fishing ceased in the maruaroa season, the month when the constellation Orion (Tautoru) appeared well above the horizon (about June).[93] Barracouta fishing commenced in the seventh month (Hakihea) – about November – and continued until April, a time which corresponds to the non-spawning period before the fish migrate to deeper offshore waters.[61, 156, 157]

Because the European settlers of the 1800s were primarily farmers, their knowledge of and interest in Māori fishing activity was limited. Settlers also had little knowledge of the fishes present in New Zealand waters and their interest was largely restricted to a few food species, and unusual or dangerous species. While praising the greater richness and flavour of familiar fishes such as John Dory, flounder, mackerel and even eels, species they recognised from 'home', the settlers sometimes described many of the new, unfamiliar species in unflattering terms: "… indeed of a large number of the New Zealand sea fish, it may be said, that they are poor in flavour and coarse in flesh, affording a most striking contrast in this respect to the many delicious species found in English waters…"[158] Crayfish were occasionally described as "very insipid" or "not highly flavoured". Groper were "…not bad by any means when one is hungry…"[159] and even snapper were described as "…dry, insipid and inferior."[3]

Mātauranga Māori and scientific knowledge based on the fauna exploited at that time, using the available fishing techniques, were not equivalent. In 1859 A.S. Thomson, a surgeon in the 58th British Regiment, noted that around 100 species of coastal fish had been named by naturalists, but the list was incomplete, considering that Māori had enumerated to him the names of many more they were in the habit of eating.[161]

The total number of fish species that can be identified with historical Māori fish names today (*ca.* 127) is less than the number recorded by European scientists at the end of the 19th century

---

[3] *West Coast Times*, 26th December 1867: 2

(*ca.* 225).[160] James Hector, Director of the Colonial Museum, noted that Māori discriminated natural objects in minute detail and every fish had a distinctive Māori name, including species that were considered insignificant by Europeans.[162] Today there are almost 300 Māori fish names that cannot be associated with a species because the species was not recognised by European science at the time when the Māori knowledge was readily available.[86, 88, 101] As the kaumātua (elders) and fishing experts passed away, their knowledge of fish identification and associated names was lost as European historians could document names only for those species they could recognise and identify themselves. The number of New Zealand fish species known to European science increased during the 20th century to over 400 by the mid-1950s,[156] and now exceeds 1300.[[163, 164] 165, 166]

Fish and other organisms are not arbitrarily divided into species by taxonomic scientists; rather, the classification reflects the objective reality that species are distinct, and do not interbreed in the wild. Experts, whether taxonomists, commercial or recreational fishermen, or tangata whenua, use different criteria to arrive at similar classifications. For example, Māori mythology surrounding the origins of fishes recognises that sharks descended from Punga (a son of Tangaroa), while frost-fish, barracouta and eels descended from his brother, Karihi.[167] This mātauranga reflects the classification in Western science which recognises sharks (Elasmobranchs or cartilaginous fishes) as related, but distinct from bony fishes (Osteichthys or teleosts).

New Zealand's freshwater fish fauna was not understood by fisheries scientists until the 1940s, and it was not until the 1960s that scientific names for the hierarchy of 27 species then recognised had been stabilised. With recent molecular studies the number of recognised freshwater species has risen to 38.[168] Despite Māori having an extensive inventory of names identifying the progress of each fish through its migratory cycle, Māori knowledge has not been systematically investigated,[169] and has been dismissed as unreliable[170] without any detailed enquiry into the usage of the names by different iwi.

Although the neglect of traditional Māori fishing knowledge can be attributed to there being no time overlap of knowledgeable fisheries scientists with Māori who possessed a comprehensive understanding of traditional fishing practices and nomenclature,[102, 171] this should not be a reason to be dismissive of mātauranga. Western science does not have a monopoly on knowledge. In a discussion of the taxonomy and relationships of Cyclostomata (hagfish and lamprey), a 2011 publication on New Zealand freshwater fishes stated that "…despite their unfamiliarity with Western science…[Māori]…perceptively recognised that lamprey and hagfish have some very fundamental similarities…"[170]

Māori classification is neither evolutionary nor species-specific, and biological scientists do not have adequate tools for investigating a knowledge base that is much more complex in its observations and organised by a different theory. Lacking appreciation, scientists have treated

FIG.10 **Giant kokopu. This species was one of the first New Zealand freshwater fishes described by European science and was illustrated by Georg Forster on Cook's second voyage while in Dusky Sound.** © Paddy Ryan/ Ryan Photographic.

hapū knowledge of the fauna with scepticism, or generally ignored it.[169, 170] In the Horowhenua region in the lower North Island, 19th century settlers described how Māori could readily identify and name eels according to their colour and morphology, and determine which of several lakes they had been caught in,[172] an achievement that cannot be replicated by freshwater biologists today.

From the 1860s New Zealand fisheries scientists determined species on the basis of morphology, but were sometimes incorrect in their identifications. Marine-hued lampreys (korokoro, *Geotria australis*) undertaking a seaward migration are so distinct from mud-dwelling lampreys that it was long thought there were two species in New Zealand.[173] Nineteenth-century scientists studying freshwater fish misidentified large, old smelt (paraki, *Retropinna retropinna*) as a separate species from their younger stages (ngaiore).[174] Common smelt also spawn at different times in different locations, and over a dozen Māori names have been recorded for smelt from different iwi.[175] In the 1940s Gerald Stockell, a fisheries inspector and taxonomist, treated smelt from different lakes as separate species;[174] however, today only two species are recognised (common smelt *Retropinna retropinna* and Stockell's smelt *R. anisodon*). Due to the confusion surrounding European identification of smelt (including misidentification with juvenile grayling, upokorokoro *Prototroctes oxyrhynchus*) it is now impossible to reconcile Māori names with the two species or their growth stages.

Within the whitebait catch, migrating juveniles of five galaxiid species or kokopu, are able to be distinguished from one another only with considerable experience, and attention to small, often highly subjective details.[176] Altogether the five whitebait species, inanga (inanga, *Galaxias maculates*), kōaro (*G. brevipinnis*), giant kokopu (kokopara, *G. argenteus*) (Fig.10), shortjawed kokopu (kōkopu, *G. postvectis*) and banded kōkopu (kōkopuruao, *G. fasciatus*), were given 20 scientific names.[171] The morphological approach to classification has often perplexed fisheries scientists and created confusion in naming species without providing reliable definitions for Māori names of freshwater fish: there are 39 unidentified Māori names for *Galaxias* species (and possibly other freshwater fishes). This list consists of inaka, tutuna, kaeaea, kōeaea, kaonge, kaore, karari, karekopu, karohi, kokakopako, kokaupara, kōrohe, kopara, kōpūtea, koukoupara, koupara, maehe, mahitahi, maitai, mata, matamata, ngorengore, pā, pāngoengoe, pāgougou, papa, pārare, pokotehe, pōrohe, puhi, rakahore, rērētawa, rōrōai, rōrōwai, rōwai, tātarāwhare, uaua, ururao, and whaharoa.[175] Claimants to the Waitangi Tribunal have argued that this confusion has delayed attention being given to habitat protection and sustained the Crown's unsupportable conviction that it was displacing Māori names with finer distinctions.[169]

Although there may be doubt regarding the correct application of Māori names to particular fish classified by Europeans, and superficially similar fish may have acquired incorrect Māori names, this information can still be used together with the present knowledge of the ecology of New Zealand fishes to determine some aspects of pre-European fishing activity.

For instance, the fact that eels were an important food resource throughout New Zealand is illustrated by the 200-plus names that were used by Māori to distinguish different life stages and varieties of the three *Anguilla* species found around New Zealand.[169]

Several coastal reef-dwelling species of fish that are reddish in colour have a variety of Māori names. European ignorance of the species and lack of familiarity with the Māori language contributed to incorrect documentation. Commissioner of Trade and Customs R.A.A. Sherrin recorded pakurakura (also used as a general term for reddish colour),[124] as the name for *Upeneichthys porosus* (red mullet), and maratea for *Lepidoperca aurantia* (orange perch).[88] By contrast, ichthyologist Gilbert Whitley recorded the same names, but for different species: pakurakura for *Bodianus vulpinus* (red pigfish); maratea for *Goniistius spectabilis* (red moki).[177]

The confusion in nomenclature is unlikely to indicate inferior Māori knowledge, as the fish species, although superficially similar in coloration, are very different in morphology, but may reflect different usage of the names by Māori in different regions. All four species are common reef-dwelling fishes found in shallow coastal waters of less than 80m depth and deeper waters.[178]

Similar confusion among Māori names has been recorded for wrasses found in shallow coastal habitats. Species within the wrasse group are quite different in coloration and morphology; however, in different regions, the names tāngahangaha or tāngāngā have each been used for spotty, banded wrasse, and scarlet wrasse.[44, 156, 179-181] Three species of superficially similar scorpaenid fishes, the red scorpionfish (*Scorpaena papillosus*), rock scorpionfish (*S. cardinalis*), and the seaperch (*Helicolenus percoides*) have each been referred to as matuawhāpuku or matuawhapuku, and various spellings of pahuiakaroa, puiawhakarua, puhaiwhakarua, pohuiakaroa and puaihakaroa.[175] The three scorpionfish species also have a confused taxonomic history which was not clarified until recently. Both seaperch and red scorpionfish have been misidentified as a single species, and both have incorrectly been referred to as *Helicolenus papillosus*.[177, 182] Again, all the species of wrasse and scorpionfish named by Māori are common reef-dwelling species found in coastal waters of less than 80m depth (as well as deeper waters), while other species of wrasse and scorpionfish found only below ~80-100m do not have recorded Māori names, indicating they were possibly unknown to Māori.[164, 178, 182]

Habitats exploited

Different fishing methods enabled Maori to harvest the abundant fish stocks found in all habitats: on tidal reefs hand-gathering, nets and spears were effective, as was line fishing and trolling from the shore and canoes at river mouths, in estuaries, along coasts and in offshore waters (Fig.11).

Fishing activities can be categorised according to the different habitat zones utilised. The first category, Land-based fishing, includes all fishing practices conducted from land without the use of watercraft; the second, Inshore fishing, includes fishing activities conducted with the use of watercraft up to a distance of about 5 km from shore; the third, Offshore fishing, includes fishing conducted at a distance over 5 km from land; and finally, Open sea fishing, was conducted out of sight of land.[183] These zones are independent of ocean depth. For example, deep-sea fishing can be 'Inshore', in areas such as Kaikōura where underwater canyons reach 1000m depth within 3 km of the coast.[184]

Although Māori had the capability to fish in deep offshore waters, and to fish for pelagic and oceanic fishes such as tuna (Scombridae), billfishes (Istiophoridae) and swordfish (Xiphiidae), such fishing could take place only under ideal weather conditions and in calm seas. Consequently, such fishing expeditions were infrequent, and unlikely to be accompanied by Europeans. This resulted in a lack of documentation of direct observational evidence. The likelihood of disposal of diagnostic head bones of large fishes at sea,[1] as well acidic soil conditions in middens which destroyed fish bones,[44] has resulted in an incomplete archaeological record, so that the full extent of offshore fishing by Māori is unknown.

Māori frequently travelled to known reefs well offshore.[24] Species such as groper and bass (moeone, *P. americanus*) were taken on reefs that were marked by lining up topographical features on land – a method which could enable a canoe to locate fishing grounds to within 5m at distances of 8-15 km from shore. Ngāi Tahu fishermen were locating fishing grounds using this method up to 48 kilometres from the coast in the 1840-60s[8, 92, 93] and the method was probably used extensively in pre-European times. On 15 February 1770 James Cook encountered four canoes some distance off the coast of Kaikōura[17, 19] where (as noted above) water depths plunge to over 1000m, but he made no comment on fishing activity. Joseph Banks noted that these canoes were further out to sea than any they had seen previously, and from the position of the *Endeavour* at noon it has been estimated that the canoes could have been up to 25-30 km from land[62] (albeit in sight of the seaward Kaikōura mountains which rise to a height of 2610m).

The evidence for fishing at depth is entirely circumstantial, as many deepwater species are frequently found in shallower coastal waters, and would have been taken opportunistically by Māori. Although many coastal reef species may be found deeper than 50m, other species found only in deeper water do not have Māori names, indicating that they too were perhaps unknown to Māori.[164] Tradition and archaeology are concerned with appreciably different aspects of prehistoric activity and the degree of overlap between can be surprisingly small;[185] however,

FIG.11 **Māori canoe offshore south-west of Mount Taranaki, with seabirds possibly feeding on discarded offal or bait. Hand-coloured lithograph by G.F. Angus, 1846. Alexander Turnbull Library, Wellington, New Zealand, PUBL-0014-02**

FIG.12 **A typical *Ruvettus* hook from Tuvalu. 287 x 120 mm. Museum of New Zealand Te Papa Tongarewa, Wellington, FE006529**

the maximum depth at which Māori fished has been estimated at between 50 and 100m, on the basis of the known ecological habitats of fish species represented in the archaeological record.[44, 186]

There is no evidence that Māori fished at depths below 100m: no species that occur only below this depth have been recorded from middens, nor do any of these deepwater species have a known historical Māori name. Today, more than one quarter of fish species found below 100m have a common name, although few are in general use, and most were coined only after the species had been described by taxonomists as commercial fishing expanded into deeper water. For example, the Māori name 'nihorota' for orange roughy (*Hoplostethus atlanticus*), a species known from depths of 800-1200m, is recent and was not used before the early 1990s. Indeed, the name orange roughy itself was coined only in the 1980s for marketing purposes, to replace the original northern Atlantic name, orange slimehead.[155]

Historians and archaeologists[37, 44] have commented on the apparent absence of specialised deepwater fishing techniques by pre-European Māori, with particular reference to oilfish (*Ruvettus pretiosus*), which were harvested in other parts of the tropical Pacific[37, 187-189] where reefs and atolls drop abruptly to great depths. Although *Ruvettus* was specifically targeted in some areas of the Pacific, fishing for the species was limited – at Tokelau and Ellice Islands it was reported as restricted to the sheltered lee sides of the islands on fine, calm nights, but the fish was only occasionally sought because of the purgative effect of the flesh.[190] Elsewhere, at Pohnpei in the Eastern Caroline Islands, Micronesia, 'oilfish' was taken only by trolling in surface waters at night, on rare occasions, as ocean-going craft were not available and boats rarely ventured beyond sheltered lagoons,[191] although it is unclear if the 'oilfish' species taken by trolling was in fact *Ruvettus*, which does not occur in surface waters and has a recorded depth range only below 100m.[192]

The evidence of '*Ruvettus*' fishing in the Pacific, based on a particular type of hook, has been disputed, as the hooks could have been made to catch fish of any one of a dozen other fish species (Fig.12). Hooks from the Society Islands have been described as "…unquestionably *Ruvettus*-hooks…"[188] although, according to other observers, *Ruvettus* was unknown to the local fishermen.[187]

It is unlikely that a specialised deepwater fishery for *Ruvettus* would have developed in New Zealand waters where the species is rare and found well offshore, and only north of Cook Strait (41°S) at depths of 100-800m.[193] The flesh of *Ruvettus* is edible in small quantities if prepared using particular cooking methods; however, the presence of non-digestible wax esters and other oily compounds makes it undesirable as a food species, and for this reason it has been banned in several countries.[194] The toxicity of the flesh would not have been eliminated by the Māori practice of drying fish for later consumption.[1, 67, 78]

The influence of modern intensive commercial factory fishing has resulted in a significant reduction in fish stocks in New Zealand waters and the elimination or reduction of populations of certain species from inshore waters in numerous areas. Many species that are taken commercially only in deepwater fisheries today were previously common in shallow water, and many northern species previously straggled much further south as higher population densities made more fish available for dispersal.

While snapper bones are found in middens in coastal Otago and Southland,[44] the species is rarely caught in those waters today. Fisheries biologists have estimated that snapper populations in some coastal regions of New Zealand have declined by 83-95 percent between 1930 and 1995 before starting to recover (Fig.13).[145, 195, 196] Today, kahawai are uncommon south of Banks Peninsula, but prior to the 1930s the species occurred in schools of countless thousands off the Otago Peninsula.[156]

In 1886 it was reported that ling were usually caught in 3-8 fathoms (6-15m) and were cast up on beaches outside Wellington Harbour after heavy gales in extraordinary profusion.[88] Sixty years later David H. Graham,

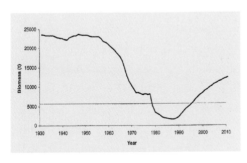

FIG.13 **Declining biomass estimate for snapper (tāmure, *Pagrus auratus*) in Tasman Bay-Golden Bay between 1930-1990. After Smith *et al*. 2003[196] The horizontal line shows the estimated biomass that supports the maximum sustainable yield.**

an ichthyologist at the Portobello Marine Fisheries Hatchery in Otago, stated that ling were common at depths of 10-25m before the 1930s and thereafter noticeably decreased in number, and he suggested that different methods of fishing were necessary.[156] A depth range of 65-100m was recorded for the species in the 1960s,[179] and by the 1980s, records suggested that ling occurred in shallow waters, but that they were most common at depths of 200-500m,[155, 178] while fishing industry websites from around 2000 state that ling is a bottom-dwelling species, taken at depths of 300-800m.

The extent of Māori fishing in offshore oceanic water has been a subject of debate.[2, 44, 164, 186] Fifteen species of tuna (Scrombridae), swordfish (Xiphiidae) and marlin or billfishes (Istiophoridae) are seasonally present in offshore oceanic waters, particularly around northern New Zealand, and several species are also found off the lower South Island. These pelagic species were probably known to Māori, although only three species, swordfish (paea, *Xiphias gladius*), black marlin (taketonga, *Makaira indica*) and striped marlin (taketetonga, *Tetrapturus audax*) have documented Māori names[175] and none has been recorded in archaeological midden deposits.[44]

Oceanic fishes are well known and common (English) names are available for 75 percent of the species, though only half are in general use. Māori names exist for one of two oceanic species of sharks, but for only about 10 percent of the oceanic teleosts, and they tend to be spectacular or unusual species that would create much interest when seen, such as marlin, flying fish (maroro, *Cypselurus lineatus*), and sunfish (rātāhuihui, *Mola mola*). Oceanic species such as swordfish were highly esteemed and were caught by Māori.

An ancient Māori proverb records that the test of a youth's manliness was the supreme task of catching a swordfish with a noosed rope: "When thou hast caught a swordfish single-handed then thou canst claim to be a man…"[197, 198] The isolation of the fishery – otherwise recorded only by the chance passing of a European sailing vessel[78] and its dependence on ideal weather conditions – probably contributed to its being overlooked by most European historians. Although not represented in the archaeological record, a single large marlin or swordfish would contribute significantly to daily food resources. Overall, oceanic fishes were unlikely to have made a major contribution to subsistence for Māori. It has been suggested that such fishing expeditions were undertaken for prestige, adventure, or simply "interesting fishing";[183] however, the desirability of large oceanic (and large offshore deep-reef fishes) to contribute to food requirements would have been a strong incentive for fishermen to travel well offshore whenever conditions permitted.

Cartilaginous skeletons of elasmobranchs (sharks, ghost-sharks, skates and rays) and bones of small fish are rarely preserved in middens, and those bones that are present are often extremely difficult to identify. An absence of fish remains in middens does not necessarily indicate an absence of fishing. Although caught extensively by early Māori,[1] elasmobranchs account for only a minor percentage of midden material (<0.9% MNI)[4] and, with the exception of elephantfish (reperepe, *Callorhinchus milii*), none have been identified to species.[44] In contrast, 21 of 27 species of elasmobranch known to occur at depths of less than 50m have a distinctive Māori name, and early settlers and travellers frequently reported thousands of sharks being prepared for storage on drying racks around coastal Māori villages, including a description of one three-tier stage which was over 250 yards [228m] in length.[25, 88]

Because of the abundance of fish around the New Zealand coast in pre-European times, it is unlikely that Māori harvesting pressure would have resulted in changes to fish populations or average fish sizes except in some limited areas. For example, the reduction in average size of crayfish in Palliser Bay between ~1150 and 1350 AD can be attributed to Māori harvesting, whereas reduction in catches of snapper in Tasman Bay between 1500-1800 AD, as indicated by a decrease of snapper bones in archaeological middens, may be as a result of climatic change.[44]

Some localised populations of inshore reef fishes, particularly labrids, show an increase in the average size of fish represented in middens over time. This may be a result of smaller specimens being preferred, but whether this was a deliberate conservation measure by Māori, or was an indirect result of selecting smaller fish (which would have been easier to dry for winter use), by using nets and traps, then switching to hook-and-line as accessible inshore reefs were depleted, is unknown.[47]

Freshwater fishing

Freshwater species provided food resources for Māori and were harvested primarily using traps (hīnaki), weirs (pā-tuna) and nets (kupenga) rather than hook and line. Fish species taken included lamprey (korokoro, *Geotria australis*), grayling, smelt, several species of galaxids or whitebait, eels, and bullies (toitoi, *Gobiomorphus* spp.), as well as invertebrates such as freshwater crayfish and mussels.[170] Although freshwater crab (pāpaka, *Amarinus lacustrus*) and shrimp (puene, *Paratya curvirostris*) were known to Māori, both were too small to be of significant food value.[109]

Historically, the freshwater grayling (upokorokoro, *Prototroctes oxyrhynchus*) was extremely common in New Zealand waters (Fig.15), even to the extent of being used as fertiliser on gardens by the cartload in the late 1800s.[176, 199] In 1869, a mill water wheel was reportedly brought to a standstill by sheer numbers of fish.[200] Grayling declined in numbers after the introduction of trout and were last seen around 1920, before being declared extinct in 1953, although the exact reasons for their demise have never been established.[174] The species was fished extensively

---

[4] Minimum Number Index (Leach 2006)

FIG.14 Small eel weir (pā-tuna) at Ngutuwera, Moumahaki River, Waitotara. After Downes 1907[87]

by Māori[1] but it has never been recorded in an archaeological midden site.[44, 170] Publications detailing results of excavations where fish bones were recorded show only large coastal species belonging to 24 families have been reported, as well as a few records of freshwater eel and kokopu at inland sites.[44, 170]

Eels, both marine (ngoio, kōiro, *Conger* spp.), and freshwater species, were a major food resource for Māori and were described as the most important fish species available.[1, 25, 38, 40, 87, 101, 104, 201, 202] Despite this, eel bones are rare in archaeological sites,[36, 44, 104] where they account for less than 2 percent of archaeological fish remains.[186] Freshwater eels have a minimum number index (MNI) of 0.33 percent for 126 sites with reliable information on fish abundance.[44] Eel bones have been reported from 12th, 14th and 16th-century archaeological sites, but not from a Waikato site where eels were traditionally known to be an important food.[36]

As there is no archaeological evidence of large-scale eeling in prehistoric times, some historians have concluded that large-scale eel fishing must be assumed to be a post-European development,[36, 44] an observation which seems unlikely, given the numerous comments on the importance of eels to Māori by early explorers and settlers.[1, 37, 87, 96, 101, 104, 109, 203] A theory of post-European development of mass harvesting of eels would suggest that construction of new eel weirs would have been seen by Europeans; however, the only observations recorded are that many existing eel weirs were removed from larger rivers by the settlers to improve navigation, were destroyed by English troops during the Land Wars, or simply fell into disrepair as alternative European food crops and domestic stock became available, and many had disappeared by the late 1800s.[7, 38, 87, 143, 172, 204, 205] [172, 209]

FIG.15 **The only known illustration of the coloration of the extinct New Zealand grayling (upokorokoro,** *Prototroctes oxyrhynchus*), **based on a fresh specimen. Watercolour by F.E. Clarke 1889. Museum of New Zealand Te Papa Tongarewa, Wellington, 1992-0035-2278 / 1**

FIG.16 **Eeling party with their catch, circa 1910. Photograph by Frederick Nelson Jones, probably in the Nelson district. Alexander Turnbull Library, Wellington, New Zealand, 1/2-28987-G**

Studies of present-day eel harvesting indicate that the negative evidence is unsound, and customs pertaining to storage, preparation and consumption of eels strongly suggest that deposition of large eel-bone middens was unlikely, even if survival of eel bones was questionable.[104] Eel bones are common in middens in other parts of the Pacific but are rare in middens in New Zealand, and it has been suggested that the absence of freshwater eel bones in New Zealand archaeological sites reflected food avoidance behaviour.[44] Eel remains in middens within the tropical Pacific may reflect differing attitudes to eels as a source of food,[44] but this theory does not address the differential survival of eel bones in middens on carbonate- or phosphate-rich coral atolls with neutral or alkaline soil pH,[206] compared to the acidic soils of New Zealand,[207] where eel (and other fish) bones may survive only in middens rich in mollusc shells that neutralise the soil pH.

Present-day eel fishing as part of Māori tradition remains sufficiently conservative to provide clear, demonstrable links with the past, and intensive eel-fishing methods, designed to take full advantage of the seasonal abundance of tunaheke (migratory eels), were developed in the prehistoric period and provided mass harvests at particular stages of the eel life cycle.[104]

386. Dried eels on racks (pataka-tuna).
The lady is Mrs Henry, of Otaki.

FIG.17 **Drying eels on racks (pātaka-tuna), at Raukawa marae during the opening ceremony of Raukawa meeting house. Photographed 14 March 1936 by George Leslie Adkin. Alexander Turnbull Library, Wellington, New Zealand, PA1-f-005-386**

An account by Thomas Brunner in 1848,[75] quoted by Augustus Hamilton,[101] describing Māori custom at Hokitika and Okarito on the West Coast of South Island, involving spiritual cleansing when dealing with eels, was used to support the theory of avoidance of eels as a food resource.[44] Elsdon Best[1] commented on the same ritual, but considered that the cleansing took place prior to setting of traps and related to eels being able to detect human scent and thus avoid the traps. Whanganui Māori distinguished a black form of eel known as tuna-tuhoro that was considered an ill omen and was never eaten,[87] and similar dark-coloured eels were also avoided in other areas of Northland and Southland,[1, 109] unlike other eels.

Non-migratory female longfin eels (tuna pouaru, *Anguilla dieffenbachii*) may reach almost two metres (6.5 feet) in length[174] and attain weights of up to 23 kg (~50 lb),[156] although eels this size are now rare as a result of commercial fishing (Fig.16). These very large eels were occasionally revered as atua or gods,[1] feared as tipua or devils,[167] or fed and tamed by some Māori. Missionary Richard Taylor[89] reported that: "…they also paid a sort of worship to an enormous kind of eel, the ruahine; to such offerings were made, by which, in the process of time, they were rendered quite tame…". Joel Samuel Polack[67] reported that large eels in a small lake at the summit of Mount Hikurangi were honoured as atua, and Elsdon Best[1, 40] noted several huge eels: one said to have occupied a pool at Te Rua-o-Puhi near Tauranga was viewed as an atua by local Māori, while others at Ruatoke and Karitane, which apparently were not regarded as atua, were eventually caught. Best stated that the custom of treating giant eels as atua was unusual and not generally followed in New Zealand.

Three species of freshwater eel are known from New Zealand, although over 200 Māori names from different regions have been recorded for different life stages, sizes or varieties.[1, 175] Colonial Museum Director James Hector[162] commented on the minute detail with which Māori named fish species, including recognising different life stages. Although a few important food species may have had up to a dozen Māori names, most have only one or two,[164, 175] and the large number of names associated with freshwater eels indicates their socio-economic importance to Māori (Fig.17). While some Māori may have had particular customs, there are only three eel varieties – arokehe, ruahine, and tuhoro – recorded as not being eaten.[208]

Eels are of particular importance among traditional Māori foods today, and pre-European Māori had a marked preference for this fish as it provided an ideal food source in the Māori

FIG.18 **Eel weir on the Whanganui River, photographed in 1921 by James McDonald. Alexander Turnbull Library, Wellington, New Zealand, 257-76-3**

hunter-gatherer subsistence economy. The taking of eels was an important activity throughout the year, with capture methods being intensified during migratory runs. Permanent weirs were constructed in many waterways but never blocked more than half of the waterway, to ensure that others downstream could harvest their share of the migratory eels. Throughout New Zealand eels were highly sought after, and specialised fishing technologies targeting eels (perhaps unsurpassed anywhere in the world) were developed, including the construction of huge weirs (pā tuna, pā auroa, pā tauremu) up to 400 yards (360m) in length (Fig.18), as well as canals, traps, holding pens and numerous hīnaki, or eel-pots.[37, 87, 101, 104]

The weirs were highly valued and protected from intruders, as mass capture of eels during the autumn migrations provided a major contribution to the prehistoric Māori diet, constituting a valuable source of fat.[1, 104, 109] The calorific food value for eels was nearly twice that obtained from freshly harvested kūmara, or fern-root, and considerably higher than any other fish species. In addition, eels exceeding 114 cm (45 inches) in length have over 17.5 percent by weight of fat, thus are the perfectly balanced food, supplying essential fatty acids, sufficient protein and necessary vitamins and minerals.[44]

Early accounts related in Māori oral histories and reported by explorers, settlers and historians indicate that mass harvesting of eels was well developed and an important part of traditional food supplies at the time of European contact.[1, 14, 24-26, 39, 40, 51, 77, 106, 107, 109, 111] In some areas Māori did not harvest eels because of ritual restrictions or beliefs.[44] Restrictions of tapu applied to eels in certain circumstances,[1, 156] some of which still apply today,[104] including cooking eels in separate ovens for ceremonial feasts at which women were not present, and ritual cleansing prior to setting traps.[1, 75, 101] The capture of eels was always celebrated with numerous religious rites and ceremonies, both before and after the return of eeling expeditions. As with marine fish, eels were never cleaned near the place where they were caught.[172, 209]

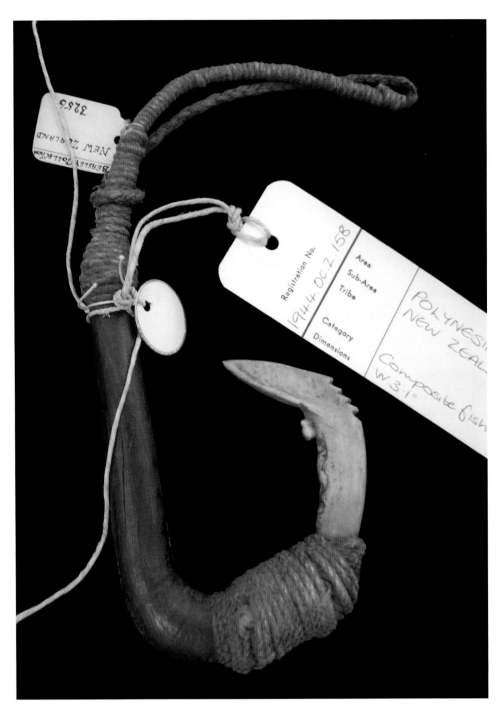

FIG.19 Wooden fish-hook with a bone point fashioned from the jawbone of a dog. 180 mm length. Date unknown.
British Museum, London, 1944.Oc.2.158

Matau: *traditional hooks*

"... their hooks, are pieces of root... to one end they attach a very sharp fish bone, the point of which bends inwards following the shape of the wood; I doubt whether they catch great quantities of fish with this implement..."

French Explorer Pottier de l'Horme, 1770 (Ollivier and Hingley 1982: 85)

European explorers were quick to note the profusion of fish in New Zealand waters and the importance of fishing to Māori. In 1769, Joseph Banks, while noting that Polynesian fishing methods were "vastly ingenious", commented on the unusual shape of Māori fish-hooks, stating that, in his view, "...their hooks are but ill made, generaly [sic] of bone or shell fastned [sic] to a piece of wood" (Fig.19).[19]

Banks emphasised that fishing with nets was more important to Māori than fishing with hooks: "...with their netts [sic] they take fish much easier than they could do with [hooks]..."[19] Johann Forster, naturalist on Cook's second voyage, mentioned the purchase of fish-hooks in Queen Charlotte Sound on 1 June 1772: "...otherwise we traded in many fishhooks; these were very unshapely, made of wood and with a piece of serrated bone affixed, which according to them [Māori] was human bone..."[21]

Other early explorers were perplexed by the odd appearance of the Māori fish-hooks when compared with their own metal hooks. Pottier de l'Horme, an officer on board de Surville's ship *St Jean Baptiste* in 1769, commented on the odd shape of the Māori hooks, expressing doubt as to their efficiency and function.[20, 66] William Anderson, ship's surgeon on the *Resolution* during Cook's third voyage in 1777, noted that Māori "...live chiefly by fishing, making use... of wooden fish-hooks pointed with bone, but so oddly made that a stranger is at loss to know how they can answer such a purpose..."[17]

Academics have stated that "as far as new fishing technologies and specialised knowledge are concerned there are few signs of significant achievements in prehistoric New Zealand."[44] However, the traditional Māori fish-hook was as efficient at catching fish as any modern steel hook and represents a design that has recently been recognised by fishermen as having superior qualities.[2] Many commercial longline fisheries have now adopted the traditional circle-hook shape in recognition of its advantages and improved catch rates (Fig.20).[210, 211] Despite this, the traditional Māori hook is still sometimes dismissed as "…typically rather clumsy…"[44, 170]

Prior to the European introduction of metals to New Zealand, indigenous Māori relied on natural materials of shell, bone, ivory, wood, and stone to make fish-hooks.[1, 140] Unlike metals, natural materials of bone, shell, wood or stone cannot be softened and bent to shape. The strength of the material is limited, and points, although stout, cannot be made with fine, sharp barbs to hold the fish. Consequently, traditional Māori hooks did not function in the same manner as modern metal J-shaped hooks.

Sharp points and barbs required for piercing and holding the fish on the line, as with present day metal hooks, could not easily be manufactured from these materials (Fig.22). Because natural materials lack the strength of metals, hook design had to be a compromise so that the hook functioned efficiently to catch fish without breaking.[2] Consequently, Māori traditional fish-hooks differed markedly in design from modern metal hooks and were often large, distinctive, and evidently highly variable in shape.

The traditionally shaped circle hook and the internal barb hook designs were developed and used by Māori to harvest the rich fisheries resources of Aotearoa. The nature of the designs, as well as their function and unrecognised benefits, were lost after the introduction and use of metals for manufacturing fish-hooks, resulting in many traditional hooks being incorrectly interpreted as ceremonial or magico-religious objects. It is perhaps ironic therefore that today many fisheries regard circle hooks as representing an advance in hook design, when it is a re-discovery of a much older technology.

Fishers of all cultures are often reluctant to share their fishing expertise, except with close family members.[125] Stylistic differences in Māori fish-hooks reflect individual preferences and are more likely to change than functional traits which may only be associated with regional and chronological variation.[212] Wooden components of artefacts, including fish-hooks, have not persisted in archaeological sites[37] except in a few waterlogged locations and one or two dry caves.[36, 213] Although the

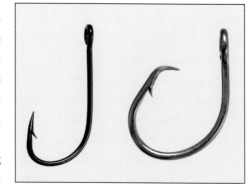

FIG.20 **Modern metal J-shaped hook (left) and circle hook (right).**

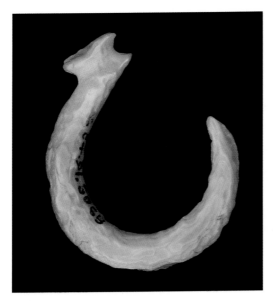

LEFT FIG.21 **Abalone shell rotating hook from Southern California. 38 x 30 mm. 1000AD Musée du quai Branly, Paris, QB 71.1884.91** RIGHT FIG.22 **One piece bone rotating hook. Date unknown. Museum of New Zealand Te Papa Tongarewa, Wellington, ME002740**

form of fishing gear may be adapted and may change with time, key functional components are retained and provide a means for tangible extrapolation into the past.[214]

The unusual design of the Māori fish-hook was not unique in its basic shape, and was predated by millennia in other Neolithic and Mesolithic cultures (Fig.21). Many museums and other institutions worldwide have collections of Neolithic fish-hooks, which are of significant interest because of the similarity of design to Māori and other Polynesian fish-hooks. This similarity has arisen from the use of natural components such as wood, shell and bone, and the associated limitations of the strength of these materials, requiring similar design solutions to produce an effective fish-hook. Neolithic bone hooks from Europe, and bone and shell hooks from the Americas in museum collections, reveal a convergence of design, often with localised variations in the form of an internal or external barb. Generally only one or two-piece bone hooks have persisted in archaeological sites. Older European Neolithic composite hooks, possibly made from wood and shell, have not been found.

Hook function

The present-day use of metals allows fish-hooks to be made in any desired shape or form – the metal can be softened by heating, then bent to shape and hardened by tempering, before being sharpened by filing. Hooks made of strong metal can be thin and slender, and are typically designed with the point of the hook orientated parallel to the shank and a reversed barb. The thin but strong shank allows bait to be threaded directly onto the hook, and the fine, sharp point which is parallel with the direction of the line penetrates the fish when 'set' by a strong upward jerk on the line by the angler using a rod as a lever. When the line is under tension it pulls in the same direction as the point of the hook, and used in conjunction with a long rod, the lever action greatly increases the force that can be applied to the hook.[2, 210] Once the fish has taken the bait and is hooked, it is held securely by the reversed barb.[210]

Māori hooks have been considered to be "impossible looking" by many observers who were unable to decide how the hooks functioned, with many recorded observations suggesting that the hooks probably didn't catch many fish.[17, 19, 66, 67, 96, 101, 139, 140] Some 230 years after the Cook voyages, when Banks and Anderson first questioned the Māori fish-hooks, a study showed that the unusual design was related to how the hooks functioned – catching fish by rotating away from the direction of the point and trapping the jaw of the fish,[2] rather than penetrating the fish in the manner of a modern metal hook or by acting as a lever.[44, 215]

Various theories had previously been proposed to explain circle-hook function. One suggested that the design improved hooking fish as they tried to expel bait they could not swallow.[216] This contradicts other observations that the inward directed point has advantages in reducing the chance of the hook snagging on the sea bottom,[125] and it is unlikely that the inturned point would facilitate hooking a fish in this manner.

A second lever-hook theory proposed that the fish hooked itself while swimming away, through the shank acting as a lever as it became free of the mouth, causing the point to penetrate behind the jawbone as the hook rotated forwards in the direction of the point, thus improving the chance of fish being hooked.[44, 215] Fisheries observers in the tropical Pacific in the early 20th century noted that the line leading away from the inner edge of the shank of a circle hook caused the hook to rotate inwards *away* from the direction of the point;[187, 217] hence, it cannot act as a lever to facilitate the point penetrating the fish.

The success of the traditional circle-hook design relies on the fishing line leading away from the inner side of the hook shank at a right angle to the point, combined with the width versus the depth of the fish's jawbone in relation to the width of the gap of the hook (the distance between the point of the hook and the shank). A fish jawbone is relatively thin but deep and could pass easily through the narrow gap of the hook, but the vertical depth of the jawbone exceeds the width of the gap. A mechanical explanation for circle hook effectiveness is based upon simple physics (Fig.23).[125]

The fishing line was attached to the hook by a snood (a short length of line permanently attached to the hook, bound with a whipping of fine twine to protect it from the teeth of the fish). The snood was lashed to the hook in a narrow, angled groove, with the line leading away from the inner side of the shank, so that when under tension the line pulled at right angles to the point of the hook.[187] As the stout bone point of the hook was required to guide the hook into position and not to penetrate the fish, it was not embedded in the bait.[2]

The bone point directs the fish's jawbone through the narrow gap between the point and the shank, into the loop of the circular hook which acts as a trap or snare to hold the fish (Fig.24). When a fish attempts to consume a baited circle hook by engulfing the bait, the jaw slips between

Anatomy of a fish-hook

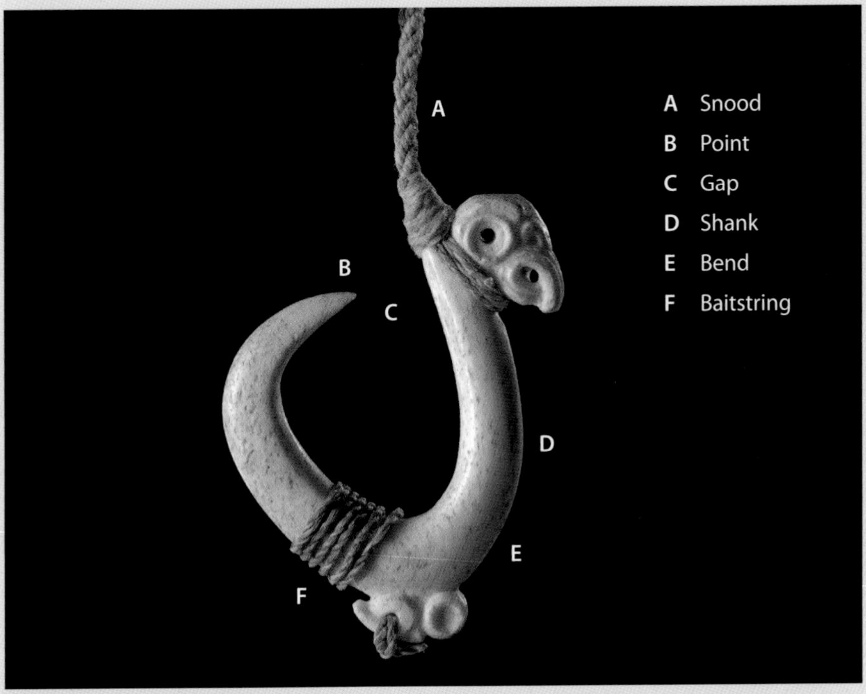

A Snood
B Point
C Gap
D Shank
E Bend
F Baitstring

As natural materials are not as strong as metals, traditional Māori fish-hooks (matau) had to be comparatively thick. Bait could not be threaded onto the hook; instead it was tied to the shank (papa-kauawhi) and bend (kou) with fine string (pakaikai). The snood (takā), tied in a groove below the head of the hook (koreke) was protected from sharp teeth by a lashing of fine string (whakamia).

The Māori fish-hook is not required to penetrate the fish in order to hold it on the line. Hence the traditional Māori fish-hook differs from modern metal hooks in having an inward directed point (mata) with a very narrow gap between it and the shank, and in the absence of a reversed barb (kāniwha). The fish was caught by the jaw slipping through the gap of the hook and was then held on the line (aho) by the snood pulling at right angles to the direction of the point. This caused the hook to rotate around the jaw and prevented it from reversing out.

FIG.23 Parts of a hook.

Rotating hook function

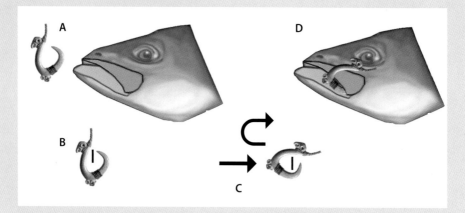

The rotating hook was attached to the line with the snood tied in an angled groove on the inner side of the shank which was at right angles to the point. Bait was tied to the bend of the hook, leaving the point uncovered (A). As the fish tried to swallow the bait, the jaw slipped through the narrow gap between the point and the shank (B). Tension on the line then caused the hook to rotate away from the direction of the point (C), and slide to the corner of the mouth, trapping the rear of the jawbone and preventing the hook from reversing out (D).

FIG.24 **Rotating hook function.**

the point and the shank of the hook; then, as the fish moves away or gentle pressure is applied by the angler, the hook is pulled to the side of the mouth. The point of the hook then catches on the jaw and pivots outwards as the pressure increases. Once tension exceeds a threshold, and the maxillary bone (upper jaw), or dentary (lower jaw) has slipped completely through the narrow gap, the hook then twists and rotates in the opposite direction to the point, and slides to the corner of the mouth where it is held by the width of the maxillary or dentary bone (now parallel to the shank), exceeding the hook's gap without the point having to penetrate the fish. When the attached line is under tension it pulls against the direction of the point so the hook cannot reverse out, thus preventing the fish from escaping.

As the hooks could not reverse out, there was no need for Māori to use rods to provide leverage to set the hook and they fished using handlines.[2] Any sudden jerk on the line before the hook traps the jaw would simply pull the hook out of the fish's mouth,[37] while relaxation of tension on the line did not allow the fish to escape, as the design of the hook prevented the hook from backing out on its own and would hold a fish even under slack line conditions.[210] Up to 10 hooks could be fished on a line, and spreaders (pekapeka) selected from saplings with radiating branches were used to keep the hooks from becoming entangled with each other.

The circle design of the rotating hook allows the fish to trap itself with no assistance required from the angler and is particularly useful in deep water or strong current where setting the hooks by jerking is difficult.[125, 218] As the point of the circle hook faces inwards, fish are usually hooked at the side of the jaw and rarely gut-hooked.

Materials used

One-piece hooks were usually made of bone or shell, and rarely stone. Throughout tropical Polynesia, shell – usually pearl-oyster (*Pinctada* spp.) – was used to manufacture complete hooks and lure shanks. During the Māori archaic and classic periods in New Zealand, shell was used to make hooks and was predominantly utilised for points of composite hooks, and to make lures such as pā kahawai. Pearl-oyster shell was not available in New Zealand and the local pāua (abalone, *Haliotis* spp.) substitute was not as strong, hence it could not be used to make large hooks.[37, 201, 219]

Only a few shells, principally Cook's turban (kāeo, *Cookia sulcata*), top shell (mitimiti, *Coelotrochus* spp.), and cat's eye (pūpū, *Lunella smaragdus*) were large and strong enough to make one-piece hooks. Consequently, shell hooks had short shanks; examples in collections rarely exceed 50 mm in length (Fig.25).

The absence of large terrestrial animals limited the source of bone material to giant moa, seals, stranded whales, dogs and human bone. Moa bone became increasingly rare as the birds were driven to extinction; however, the abundant whale populations around the coasts provided a steady supply of bone from beach strandings, with suitable material transported throughout the country.[62]

Bones of deep-diving marine mammals (particularly sperm whale) are less dense than those of terrestrial animals,[220, 221] and frequently the only whale bone suitable for manufacturing hooks was the 'pan bone' (the dense articulating surface of the rear portion of the lower jaw), and the articulating surfaces of the atlas vertebrae and rear of the skull.

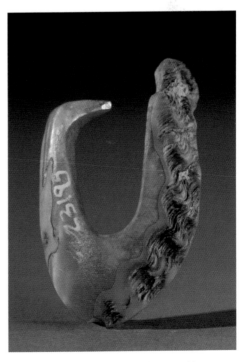

FIG.25 **One-piece fish-hook made from Cook's turban shell (kāeo, *Cookia sulcata*). 40.5 mm length. Date unknown. Museum of New Zealand Te Papa Tongarewa, Wellington, ME023195**

In contrast, the rostral (snout) bone of some beaked whales (*Mesoplodon* spp.) has been shown to be among the densest bone of any animal.[222] However, until DNA analyses can be used to identify source materials, it is impossible to determine if beaked whale rostrum material was utilised by Māori.

Seal and dog limb bones, although similar in strength to human bone,[220] were generally not large enough for hooks; however, dog jaw-bones were often used to make hook points. Human bone was prized as a material for fish-hooks because of its density and strength. The bone of enemies was doubly valuable and its use was considered an act of revenge. Hooks made from human bone were sometimes given special names, and at times this led to prolonged tribal warfare.[1]

The use of a single piece of bone limited the size of a hook that could be made; one-

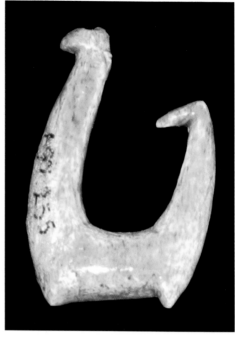

FIG.26 **One-piece bone rotating hook. Date unknown. 31 mm length. Puke Ariki, New Plymouth, A97-255**

piece bone hooks rarely exceeded 75-80 mm in length (Fig.26). The maximum size of one-piece bone and shell hooks is determined by the strength of the material required to land a large fish. Although bone from giant moa and stranded whales could be used to make larger hooks, such hooks could easily snap when subjected to the stress of a fish pulling the point limb of the hook against the line attached to the shank limb.

There are numerous archaeological examples of broken hook shanks in museum collections. The broadly rounded curve of the circle hook spread the load, and thick, short shanks reduced the chance of the hook breaking; hence, bone hooks made to a circle design were stronger than hooks with a narrow angle that allowed tension to be focused at the bend, or those with long shanks that increased the leverage a fish could apply to the hook.[212]

Large, strong bone hooks could be made in two sections by lashing a point directly to a bone shank. Although this method resulted in a stronger hook, the straight shank did not allow the point to be directed inwards as required for the circle design of traditional hooks, and two-piece bone hooks were generally restricted to lures or occasionally jabbing hooks.[2, 223]

Some studies of Neolithic artefacts have suggested that green, unburnt bone is difficult to modify, and that weathered, dessicated bone, or bone from cooked animals, is best for making tools while others consider that fresh bone is easiest to modify, being less brittle.[224-227] Soaking or boiling dessicated bone does not reduce brittleness; however, excessive heating will make the bone even more brittle.[224, 228] Any treatment of bone or preference for dried or green (fresh) bone by Māori prior to carving is unclear and has not been documented.

Because of the inherent weakness of natural materials, hooks were comparatively thick in contrast to modern steel hooks. This thickness made it difficult to thread bait onto the hook, and bait was tied to the lower bend of the hook by means of a thin string (pakaikai) attached to a small hole or lug at the base of the hook. The point of the hook was left bare, free of any bait, to ensure the hook functioned efficiently.

Crayfish, which were readily available in intertidal pools and subtidal channels, were preferred bait for most fishing, and were also ground up and used as berley (tāruru – pulped or finely minced groundbait used to attract fish).

Octopus (wheke *Octopus* spp.) was also popular for bait, but was often scarce and difficult to obtain. Shellfish including mussels (kuku *Perna canaliculus*) and pipi (*Paphies australis*) were frequently used (pipi especially when targeting tarakihi), and were sometimes par-boiled to toughen the flesh so that it remained on the hook.[8, 56] Other bait included small fishes, particularly mullet (kanae *Mugil cephalus* and aua *Aldrichetta forsteri*), which could be split and tied to either side of the bend and point-limb of the hook.

Hook characteristics such as shape, thickness and barbs are morphological traits that affect functional use, hence they are subject to selection and modification through daily use and are similar over wide geographical areas and through time. Only individual fish-hook makers' preferences and stylistic traits such as ornamentation can be used to establish temporal markers and demonstrate changes in material culture.[212]

Although overall design preferences (for example, the use of jabbing or circle hooks) may reflect changes in fishing practices and strategies over time,[218] academic use of hook shape and raw materials to illustrate cultural affinities may result in ambiguous conclusions, as artefact similarities may also result from adaptations in hook design arising independently in response to similar environmental conditions, including the use of similar natural materials, and targeting of unrelated fish species that show similar ecological habits.[212] In contrast, carved shanks, and snood attachment knobs or grooves, show stylistic change over time,[229] suggesting that specific types may reflect patterns of interaction and ancestral relationships.[212]

Museum collections have numerous examples of archaeological Māori fish-hooks made from bone and shell that have been used to provide chronological sequences and analyses of Māori fisheries as part of the historical and archaeological record. While the distinction between the earlier and later styles relies largely on 18th and 19th century evidence and undated pieces to construct a hypothetical sequence of change,[36] insufficient stratigraphic evidence is available from archaeological sites in New Zealand to convincingly demonstrate changes in hook form through time.[44]

Changes over time that have been identified include the increased use of double internal barbed hooks and barbed lure points.[223] Hooks found in the northern North Island made with bone points ingeniously sourced from dog jaws were assumed to be a late addition to Māori fishing equipment,[230] but similar points are known from throughout New Zealand as far south as Otago and have been attributed to the early period of Polynesian settlement.[201] It has been suggested that more recent classic period (or traditional phase) hooks were first developed in northern regions, and appear to have been introduced to southern regions around the mid-15th century,[36, 201, 231] possibly by Ngāi Tahu from the East Coast region of the North Island.[223] These classic period hooks were typically manufactured with wooden shanks and bone points with the component pieces held together by lashings of muka (flax); plain one-piece hooks were replaced by hooks that were adorned with bait knobs, or snood attachment knobs (koreke) often shaped as human heads.

Types of hooks

Māori crafted two principal types of hook: matau – C-shaped or U-shaped suspended bait hooks – were designed for catching demersal and bottom-dwelling fish, while unbaited trolling lures – either pā made with shell, bone or stone shanks, or pohau mangā made of wood – were for catching predatory pelagic fishes. In addition, stone and shell 'gorges' were occasionally used (see Fig.34).[37, 101, 140]

Circle (rotating) hooks

Most Māori hooks were manufactured with a strongly incurved point and are known as C-shaped or circle hooks (Fig.27). At Houhora, a 14th century Māori village archaeological site in Northland, 84 percent of hooks recovered were circle hooks.[219] Circle hooks were deliberately made with the shank leading to broad circular loop,[37] allowing the point to be directed inwards with the tip close to the shank, leaving only a narrow gap.[218, 219, 232] Elsdon Best reported a wooden circular hook of nine inches (22.8 cm) diameter, with a gap of only 1¾ inches (4.5 cm) opposite the point,[1] while many examples of one-piece bone hooks between 35-50 mm in length have double internal barbs with gaps of only ~2 mm.[44, 233]

In 1927 Clark Wissler (Curator of Anthropology at the American Museum of Natural History) had described small metal "ring-like" hooks from Japan that were similar in design to shell hooks used throughout the Pacific, which were used to take fish without injury to the animal.[234] In 1949 Sir Peter Buck noted that professional fishermen in Hawai'i had metal hooks

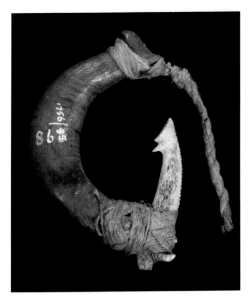

LEFT FIG.27 **Composite wood and bone circle hook. Date unknown. Auckland Institute and Museum, Auckland, 1756-85** RIGHT FIG.28 **Reverse-eye hook: the eye angle causes the fishing line to force the hook forward under tension.**

made with barbless, incurved points, based on the Polynesian design.[37] Since the 1960s many commercial oceanic and deepwater longline fisheries have abandoned the use of J-shaped hooks and switched to patterns based on the circle hook. The circle hook has also become increasingly popular in recent years in commercial and recreational fisheries because of increased catch rates associated with the design, and advantages of holding live uninjured fish on the line until they can be retrieved.[210, 211]

In reviews of fish-hook design and effectiveness, it was found that metal J-shaped hooks catch fish more readily than circle hooks, but once fish are hooked, circle hooks are responsible for higher landing rates,[210, 211, 235, 236] and are promoted as a means of improving selectivity to reduce bycatch. The advantage of the circle-hook design for retaining live hooked fish in passive fishing situations – such as longlining – is the main reason for modern commercial fisheries adopting the design. Recreational 'catch-and-release' fisheries favour the circle-hook design because of perceived conservation advantages in which the hook facilitates catching the fish by the jaw (rather than being 'gut-hooked'), thus reducing mortality and making it easier to release a fish with minimal handling. Many recreational fisheries in North America now require the use of these hooks in catch-and-release fisheries.[211]

Whilst reduced mortality was found to be true for many species, the use of circle hooks in some freshwater fisheries increased injury, thought to be associated with the feeding strategies of different species.[210] However, increased injury may be the result of poor hook design. The angle of the eye of the hook to the shank is critical in causing injury to the fish: a reverse-eye hook (Fig.28) will be driven forward under tension, often penetrating the fish and exiting through the eye, and will have minimal conservation benefit. The traditional rotating fish-hook is always attached to the fishing line with the snood leading away from the inner edge of the hook shank, at right angles the point. This causes the hook to rotate away from the direction of the point under tension.

As stated above, Māori rarely used rods, except when fishing with J or U-shaped jabbing hooks (or lures such as pā kahawai or pohau mangā lures as noted below), and usually fished with baited rotating C-shaped hooks on handlines. Rod fishing, or tihengi, when used with rotating hooks, served only to enable several lines to be fished in close proximity on rods of different lengths to prevent lines becoming tangled, and fish were retrieved by hand once hooked, while the rod remained fixed in place.[37, 103] Studies of fishing practices in Polynesia have shown that rotating hooks are particularly useful in deep water, in strong currents where setting the hook by jabbing or jerking is difficult, or in shallow water where the prey size is large.

Turtle shell was often used for making hooks in the tropical Pacific. Because the shell was slightly flexible, the fish's jaw could easily slip out if the gap widened; hence, turtle-shell hooks were made with barbs to offset this tendency.[125, 212, 218]

Modern J-hooks manufactured from metals are designed with a similar shape at all sizes. In contrast, small traditional matau are very different in shape to larger matau. Large wooden and medium-sized matau were usually composite hooks with bone points, or one-piece hooks made of wood or bone, that had a long shank with a broad circular loop, while small matau were often made with the shank and point parallel and of equal length, with double internal barbs.

U-shaped hooks

J or U-shaped jabbing hooks (Fig.29) required the use of short rods. Jabbing hooks were made in a variety of forms in order to target different fish species. These hooks were fished with a short line and a rod was used to keep tension on the line and flick the fish out of the water, in a manner similar to that used when trolling with lures such as pā kahawai and pohau mangā.[37] At the Palau Islands in the tropical western Pacific, historians documented up to nine basic types of jabbing hooks that were in use until well into the 20th century.[125]

While variation in jabbing hooks has been reported from archaeological assemblages in New Zealand,[219, 230] it is not known if this variation is related to targeting of different fish species. Jabbing hooks did not rotate and the line was attached to the shank parallel to the direction of the point for maximum efficiency. J and U-shaped hooks were much less common than C-shaped rotating circle hooks, but were particularly abundant at the Chatham Islands where they appear to have been the preferred design.

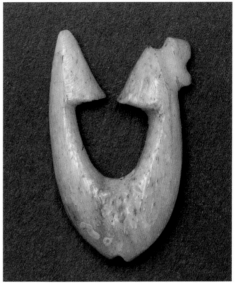

LEFT FIG.29 **Whale-bone U-shaped jabbing hook. 67 x 50 mm. Date unknown. Museum of New Zealand Te Papa Tongarewa, Wellington, ME013790** RIGHT FIG.30 **Internal-barb hook. Gifted by Lord St Oswald, 1912. Possibly of Cook voyage origin. Museum of New Zealand Te Papa Tongarewa, Wellington, ME02499**

Jabbing hooks are more commonly used in shallow water where tension can be maintained on the line once a bite is felt, allowing the angler to retrieve the fish.[212] Small one-piece U-shaped matau may also have been fished as a jig without bait, or perhaps with a small tuft of feathers. It is likely that these traditional hooks were also fished in a horizontal position, in a method of fishing that was observed and documented in the Society Islands in the early 20th century when Polynesians were still using traditional hooks.[187]

Internal-barb hooks

Internal-barb hooks (also known as shank-barb hooks) which rarely exceeded 30-50 mm in length, have two bluntly pointed barbs (kaniwha) to create a narrow gap (Fig.30), and perform the same function as the main inturned point on the larger circle-hook.[2] The double internal-barb was an integral part of this hook design and was not used as a cleat to hold bait,[37] nor was it simply a convenient way of narrowing the gap during manufacture of the hook.[44]

The point of the small hook was not directed inwards as with the larger one-piece bone or composite wooden hooks, but was directed forward, and served to guide the hook into position where the inner barbs could trap the fish. As with the larger hook design, the hook rotated when pivoted by tension on the line pulling in a direction away from the direction of the outer internal-barb.

The narrow-gap 'internal-barb' and C-shaped hook designs are common patterns found in

Neolithic sites throughout early Polynesia, Asia, the Americas, northern Africa and Europe. One-piece bone hooks with double internal-barbs have been reported from archaeological sites at ancient lake sites in the Sahara Desert dated at around 9000 years BP,[237, 238] as well as from Norway (dated at 4000-2500 years BP),[239] and similar shaped shell and bone hooks from the Americas have been dated around 1000 years BP.[142]

While it has been suggested that the use of the design across the Pacific may represent contact between different cultures,[188, 240] it is more likely that the similarity has arisen independently, as these design requirements are needed to deal with inherent structural weaknesses when making hooks from common natural materials such as bone and shell. Composite wooden hooks from older Neolithic cultures have not survived archaeologically and are unknown.

Hooks with internal barbs are more commonly represented among archaeological hooks recovered from sheltered eastern bays and north-eastern coasts of New Zealand (e.g. Northland to Bay of Plenty; Hawke Bay; Golden Bay, Nelson), although there is a paucity of archaeological research on the west coasts of both islands.[201, 233, 241] These hooks were possibly used in a unique manner to catch large benthic fish, where the narrow gap functioned by catching elements of the branchial (gill) arch in a similar way to the larger rotating hook's gap that functioned as a trap to hold the jawbone.[242]

Elsdon Best recorded an ancient karakia or incantation that was chanted over hooks before fishing to ensure that the hooks functioned effectively.[1]

Na, mo te matau hi ika tenei karakia

Tenei au he atu [?au tu], he au noho ki nga tipua aro nui, aro rangi, aro nuku

Ki tenei taura nga tipua, na nga tawhito nuku, nga tawhito rangi

Ki enei matau riki, ki enei matau piha, ki enei matau pakiwaha

Kia tau aro, kia tau whiwhia, kia tau rawea mai

Kia piri mai ki tenei tama, kia rawea mai ki tenei tamaroa

He awhi tu, he awhi noho taumanu kia tamaua take

Kia tamaua piri, kia kai nguha, kia kai aro, kia kai apuapu

Kia taketake nui, kia taketake aro ki enei matau

Hirere awa, hirere au, hirere moana

Tamaua, tamaua take, eke eke uta ki runga I taku waka…e

Hui…e! Taiki…e!

This is an incantation for a fishing hook

This is the current that connects us to the elements from the heavens above to the earth below

Bound here by the ancient elements,

Bless these small hooks, these hooks for the gills, these hooks for the mouth

That they strike true, that they are well fastened, that they are wrapped well

That they become one with this fisherman,

Embracing the line, held firm in the canoe, holding fast

Holdfast hooks, hunt your prey, strike true, hook me many fish

Be long lasting my hooks

From the rivers, from the currents, to the ocean

Hold firm, board this vessel, journey with me

We are united!

Two phrases within the karakia are of particular interest and provide clues to how some of the double-internal barb fishing hooks functioned.[242]

Kia tau whiwhia, kia tau rawea mai...
That they are well fastened, that they are wrapped well...

Circle hooks, with bluntly pointed tips, captured and held fish by trapping the upper or lower jaw as they rotated away from the direction of the point and held the fish on the line without penetrating the flesh – a process that could be described as 'wrapped well'.

The second phrase is intriguing, as it suggests a unique method of fishing related to small internal-barb hooks:

Ki ēnei matau riki, ki ēnei matau piha...
To these small hooks, these hooks for the gills...

Fish feeding on the sea floor will often suck in quantities of sand and shell debris along with food items by suddenly expanding the gill covers. Sand, small shell fragments and detritus can then be ejected between the branchial arches and out through the gill opening. The anterior or leading edge of each gill arch is lined with comb-like structures known as gill rakers.[243] When the mouth of the fish is closed, the gill rakers lie flat along the gill arch, but as the fish expands the branchial cavity by opening the operculum (gill cover) to expel water and debris, the gill arches flare outwards and the rakers become erect, forming a grid that allows water and detritus to pass while preventing larger food items from being ejected, which can then be swallowed.

Small to medium-sized pelagic feeding fish have long, closely spaced gill rakers. While it is possible that a hook could catch the branchial arch as unwanted material is ejected out through the gill opening,[233] even a small hook could not pass through the mesh or grid created by the gill rakers except when targeting large, benthic feeding fish.

Internal-barb hooks with original attached lines collected by James Cook and other early European explorers are notable in that the lines are of exceptional thickness in relation to the size of the hook (Fig.31). These lines are often as thick as or even thicker than lines attached to much larger wood-shanked rotating hooks.

New Zealand flax provided fibrous material for fishing lines and was recognised in the early 1800s as equal to or superior in quality to the jute, hemp and sisal in use by Europeans at the time.[17, 67] It is possible that very large benthic or bottom-dwelling fish may have been the targeted species using the internal-barb hooks and heavy fishing lines. The relatively wide-spaced gill rakers of very large (>1 metre) benthic fish, such as ling, groper and bass, would enable the small hook to slip between the gill rakers and catch on the branchial arch. The occurrence of the internal barb hook in archaeological sites along the eastern coast and

FIG.31 **Internal barb hook with dressed flax or muka line collected during the Cook voyages.** Georg-August-Universität Göttingen, Oz 329

large sheltered bays reflects areas where it was possible to regularly fish from canoes well offshore and target large reef-dwelling fishes at depth. Such reefs were not as accessible on the exposed west coast where larger waves prevented safe fishing at distances from shore.

Studies of fishing practices at Palau have noted that there is a functional relationship between the gap size of rotating hooks and fishing depth – hooks with the narrowest gaps were used at the greatest depths where fish were often large and of high quality.[125] Evaluation of this relationship between hook form and fish catch in the New Zealand archaeological record may help in interpreting pre-European fishing activity;[212] however, large species such as groper, bass and ling are not represented in the archaeological record in great numbers.[44] Heads of large fish were disposed of at sea as an offering to the god Maru,[1, 26] a practice that would result in a lack of diagnostic head bones in middens.

In his book *A Treasury of New Zealand Fishes*,[156] David Graham reported that in 1900-05 fishermen could hook two to three dozen groper per hour off the Otago Peninsula, and 1922-

27 two men working could catch 5 to 15 dozen 80 lb (36 kg) fish per day. A single fish of this size would have provided Māori with significantly more food than much smaller species such as barracouta, blue cod, snapper and others that are more numerous in middens.

The capture of large benthic reef-dwelling fishes by using small hooks to entangle the branchial gill arch, recorded in mātauranga, represents a fishing method that is unknown today but has been documented in studies of fishing methods.[242, 244] In an introduction to a paper on oilfish (*Ruvettus pretiosus*) fishing in the Pacific,[188] Clark Wissler, Curator of Anthropology at the American Museum of Natural History, noted in passing that "…scattered over the Pacific are small shell hooks resembling an open ring, which… seize the gills of fish and hold him firmly, but without injury…" Wissler noted that similar hooks were known from Japan and the American states of Alaska and Washington,[188] and comparable bone hooks are also known from Chile.[142] Wissler suggested that the distribution of the hook illustrated a common centre of dispersal of culture; however, this dispersal theory has since largely been discredited.[245] It is likely that hook design converged in different cultures as a result of similar raw materials being used to target fish in ecological habitats that shared common features.

The re-discovery of the circle hook is regarded as a design breakthrough that enabled improved landing rates in longline pelagic and deepwater fisheries, and as fish are rarely harmed by being gut-hooked, the design has also been regarded as innovative for recreational catch-and-release fisheries.[210, 211] The use of small gill-hooks by Māori also represents a previously unrecognised technological achievement that was widespread in Polynesia and the Americas. The removal of large benthic and demersal fish species from coastal waters by intensive commercial fishing in the 20th century[146, 156, 246] may make experimentation with this fish-hook design and confirmation of its effectiveness problematic.

Composite hooks

Although early Europeans noted that fishing with nets was more important than hook-and-line fishing,[17-20, 23, 24, 26, 66] stone and bone portions of fish-hooks survive well in archaeological sites while wooden components, flax lashings and nets do not.[36, 37, 201, 219, 223, 231] Flax components of hīnaki (wickerwork traps) lasted only about a month, while other materials such as kiekie would last for five to seven years.[1]

Early fish-hook types reflect initial attempts to copy Polynesian prototypes in local materials, and subsequent adaptations to suit local fishing conditions were reflected in a general trend to greater ornamentation. There were also regional and other variations attributable to individual hook-maker preferences.[26, 36, 37, 44] As with other Polynesian cultures, it is likely that Māori fishermen were reluctant to share their successful hook-design features; mātauranga surrounding fishing activity was carefully guarded and often not shared even within hapū or iwi.[123, 247]

FIG.32 **Composite hook with whale-bone point. Date unknown. Auckland War Memorial Museum, Auckland, AWM 14626.5**

Archaeological studies have shown that early Māori fish-hooks (ascribed to the period of Māori culture referred to as archaic or settlement phase), were predominantly made from one piece of bone,[201] and lures were generally manufactured with stone shanks (minnow lures), similar to shell lures from Eastern Polynesia, with simple non-barbed bone-points.[37, 223, 248-250] Wooden archaic hooks are rare. Wooden hooks or lures with flax lashings (Fig.32), as with flax nets, are less likely to survive well in archaeological sites.[201, 223, 231] In contrast, bone barracouta or minnow-shank points and composite bait hook points, which were associated with wooden shanks, are well represented.[223, 231, 251, 252] Even old wooden hafted tools such as adzes are rare archaeologically as the wooden hafts and fibrous lashing material decayed, leaving only the stone heads.[37, 253]

Māori trained growing plants and young saplings into the desired curve for wooden shanks of hooks, then harvested them after they had grown and become rigid. The notable traveller and botanist, William Colenso, described how branches of cottonwood (tauhinu, *Ozothamnus leptophyllus*), and climbing fern (mangemange, *Lygodium articulatum*), were grown into suitable form (Fig.33), then hardened by heating in hot earth beneath a fire to toughen the wood.[24] Fresh growth on certain trees, including celery pine (tanekaha, *Phyllocladus trichomanoides*), could be trained into a hook-shaped form for harvesting at a later date.[1, 26, 101] A cache of fish-hooks found in a small rock shelter at the Manukau Heads in northern New Zealand, was found to include composite hooks, some of which had been manufactured using shanks from pohutukawa (*Metrosideros excelsa*) root that had been hardened by fire. The hooks in this cache were fashioned with points of whale and albatross bone, and in two instances, ostrich foot shell (pūtara, *Struthiolaria papulosa*).[250]

The bone or shell point of a composite fish-hook was usually lashed to the wooden shank with muka, fibre prepared from New Zealand flax or occasionally kiekie fibre. Coarser fibre from the cabbage tree was rarely used for delicate hook lashings, but was commonly used for ropes associated with fishing nets.[1] Throughout tropical Polynesia, fibre prepared from coconut husk

FIG.33 Composite hook with a strong curved wooden shank, made by training a growing plant to the desired shape, lashed to a bone point. 128 x 99 mm. Date unknown. Museum of New Zealand Te Papa Tongarewa, Wellington, ME 014838

(*Cocos nucifera*), known as sennet or afa, was preferred for lashing fish-hooks, while nets were made from the fibre of parau (*Hibiscus tiliaceus*) or alongā (*Pipturus argenteus*).[22, 254] Coconut, parau and alongā fibres were not available in New Zealand.

Although Māori reportedly sometimes used gum from native plants such as rangiora (pukapuka, *Brachyglottis repanda*) to preserve lashings, only a few of the composite hooks collected during the Cook expeditions or other European voyages of discovery appear to have been treated with resin. In some instances (particularly the Hunterian Museum collection), resin may have been applied by collectors.[249]

Wooden hooks were deliberately made large in relation to the size of the fish targeted, and were used to catch fish that could not swallow the entire hook. The preferred size of hooks rarely exceeded ~120 mm,[86] as the hook was only required to be large enough to slide over and secure the fish's jaw. The composite hook functioned in the same manner as the smaller C-shaped bone hooks, and the point of the hook acted as a guide, directing the fish's jaw down the shank of the hook until it reached a position at which the distance between the point and the shank narrowed significantly. The point of the hook was not baited, which enabled it to guide the fish's jaw into the trap formed by the loop of the hook, and form an obstruction to prevent its sliding back out. These points were serrated but usually did not have large reversed barbs as found on modern metal hooks.

Hooks for large deep-reef fishes such as groper, bass, ling, sharks and other large species needed to be more robust. The hooks were composite and made with differing materials such as a strong, curved wooden shank lashed firmly with a stout bone, shell or stone point. Māori considered the shape of the hook most important and hooks preferred for catching even large sharks were short in the shank, never exceeding the breadth of three fingers, "the standard measure".[86]

Very large sharks (~3m or more in length), particularly mako (mako, *Isurus oxyrhinchus*) and to a lesser extent, white shark (mango-taniwha, *Carcharodon carcharias*), were not popular as food, but were sought after for their teeth which were worn as ear pendants to indicate

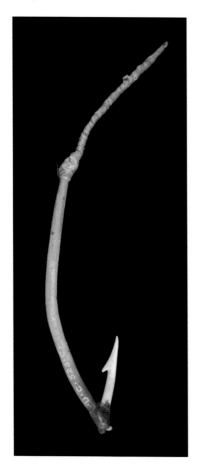

FIG.34 **Composite seabird hook. Date unknown. National Museum of Scotland, Edinburgh, A.UC523.C**

chiefly rank. To avoid damaging the teeth (and because hooks that were strong enough could not easily be made), large sharks were rarely taken by hook and line. An elderly Māori chief described the method of catching these sharks on the east coast to William Colenso in the 1840s (ko tona hii tonu tenei o tenei ika o te mako):[25, 86] sharks attracted by fishing activity would often approach canoes at sea. As the shark neared the canoe, a large bait would be lowered quickly, tempting the shark to dive after it. This caused the shark's tail to lift out of the water and a strong noosed rope would then be tossed over the tail, hauled tight, and held until the fish was exhausted. These large sharks where rarely eaten, but the heads were removed at sea using a saw made from the serrated teeth of a six-gill shark (tatera, *Hexanchus griseus*) set into a wooden blade, and the body left to drift away.[25, 101]

FIG.35 **Polynesian fishing gorges. 82 x 4 mm. Date unknown. Museum of New Zealand Te Papa Tongarewa, Wellington, FE000471**

Large hooks made from wood probably floated and were fished with a stone sinker: the wooden hook would float into a position where the heavier bone point, and the shank (to which the weighted line was attached), were directed downwards (rather than upwards as with steel hooks). Beasley (1928: plate 25)[140] illustrated an example of a hook with an inverted figure carved into the shank, which would sit upright when the hook floated, attached to a weighted line. This floating position would help to increase the rotation of the hook when tension was applied to the line from above.

Although it has been suggested that the presence of large hooks strongly indicates shark fishing,[219] other species of large fish (~1m or more in length), such as groper, bass, and ling, were available to Māori in shallow coastal waters at depths of 30-70m or less,[88, 146, 156, 164] and were targeted using these large hooks.

Many museum collections have examples of composite hooks that are unusually slender (Fig.34) and do not appear to be strong enough to hold a fish of sufficient size that could take the hook. These slender hooks are found in a variety of forms, including many post-European contact examples with metal points, and are often characterised by stout shanks, wide circular gaps between the point and shank, and with points that are not directed inwards.

Other examples have long slender snood shanks with a bone point at the tip, or are distinctive in having two slender shanks made from a forked branch which has been tied together so they are almost parallel and of equal length, with recurved points fashioned from shell or bone.[250] Examples from the East Cape region are common in museum collections. It has been reported that these hooks were possibly used for catching seabirds rather than fish.[1, 26, 101] If this is the case, the number of these slender composite hooks in museum collections suggests that hook-and-line was an important method for taking petrels (Procellariidae) and other seabirds such as albatross (Diomedeidae) by Māori, for both food and feathers.

Gorges

Some archaeological sites in New Zealand have yielded bone artefacts known as gorges (Fig.35). Gorges were a device used by many Neolithic cultures, and are frequently found in prehistoric sites worldwide.[210] A gorge was made from a slender stone, bone, or portion of shell, usually 50-100 mm long, and was attached to a line, which was knotted through a hole in the centre. The fish swallowed the gorge (hidden inside a bait), end first, and held the fish by becoming trapped between the gill rakers, or tension on the line levered the gorge across the fish's throat, trapping it in place. There are drawbacks to fishing with a gorge as it is hard to conceal, difficult

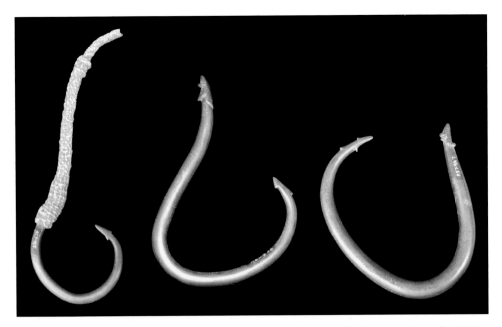

FIG.36 **Circle hooks made using copper ship's nails. Dates unknown. Puke Ariki, New Plymouth, from left: A57-955, A57-955, A80-536**

to bait, hard to hook large fish on, and liable to lose its hold while the fish is being played. Although examples have been reported from archaeological sites in New Zealand, Elsdon Best noted that their use was unknown among his Māori contacts.[1]

Metal hooks

Māori quickly recognised the superiority of metal over natural materials for tools following the arrival of European explorers (Fig.36). James Cook gifted metal tools to local Māori and traded these for supplies of fish, often providing metal nails in exchange. At Hawke's Bay, a few days after sighting New Zealand in late October 1769, Cook gave local Māori gifts of linen cloth, trinkets and spike nails, and noted in his journal that they placed no

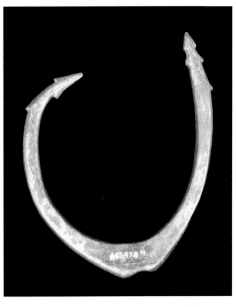

FIG.37 **Fish-hook made using a horseshoe. Date unknown. Puke Ariki, New Plymouth, 7-974 068**

value on the metal nails. A few weeks later, in January 1770 off Cape Palliser, several hundred kilometres south of Hawke's Bay, the crews of three Māori canoes boarded the *Endeavour* and requested nails, which they had heard of but not seen.[42] Metal hooks were avidly sought after by Māori, and many later explorers, sealers and whalers often used metal fish-hooks as a form of currency.[23, 82]

Metals, including copper ship's nails, wire, and even horseshoes (Fig.37), were used by Māori to make fish-hooks. The hooks were carefully fashioned following traditional circle designs, with barbless, inturned points and were quite distinctive compared with the mass-produced J-shaped metal hooks introduced by Europeans. Māori continued to make metal hooks following the traditional circle design, until cheap mass-produced, European J-shaped steel hooks became widely available.

4

Pā: *trolling lures*

"The native method of catching this fish is singular, and worthy of notice. A hook made of bone, with a piece of the glistening shell of the 'mutton fish' attached to it, but without any farther bait, is trailed at a short distance behind a canoe, which is being paddled with the greatest possible speed. The resemblance of the piece of shell, in its rapid motion along the surface, to a small fish, causes the salmon to seize it with great avidity, and immense numbers are thus caught."

Artist and explorer Charles Heaphy, 1842 (Heaphy 1842: 49)

Trolling lures, known as pā (Fig.38), were used by Māori to catch pelagic or midwater feeding species, such as kahawai and barracouta. Pā were generally made using a wood, bone, shell, or stone shank, with a short sharpened shell, ivory or bone point set at the distal end.[23, 26, 82, 140, 249, 252, 255-257]

Several different types of lures were used. These included pohau mangā or barracouta lures, manufactured with a long wooden shank; minnow lures made with stone shanks and bone or shell points; pā with bone or shell shanks and points; pā with the shank and point made entirely from the rim of a pāua shell; wood-backed pā kahawai with pāua shell lining a wooden shank; and unusually slender lures made with a pāua shell shank and long slender wooden or dogfish-spine points.

Most lures had tufts of feathers from kingfisher (kotare, *Halcyon sancta*), blue penguin (kororā, *Eudyptula minor*), or occasionally kiwi (*Apteryx* spp.), at the distal

FIG.38 **Pā or trolling lures illustrated by Sydney Parkinson on the first of Cook's voyages. Detail from Parkinson 1784, Pl. XXVI**

end to attract fish. Trolling lures were not baited and were fished by being towed through the water behind a canoe, or were fished by hand at river mouths. Occasionally a wooden implement known as a reti was used to troll lures along a beach – the reti acted in a similar fashion to an otter board and kept the lure well out from the shore, although it is unclear if this method was known to pre-European Māori.[1] Lures made using the curved pāua shell would spin when trolled through the water, reflecting light from the shell and attracting fish.

Pā

Straight-shank lures, including those from New Zealand and elsewhere in Polynesia, were designed with the line extending to the base of the shank to hold and support the point of the hook: this shank-line or snood is a necessary component to secure the point when playing a large fish (Fig.39).

Pāua shell pā kahawai lures rely on the point being secured by firmly lashing it to a bulge at the base of the shank. It was not possible to attach the trolling line directly to the base of the point of pā kahawai as the concave pāua shell prevented the line from running down the inner face of the shank. Points of curved pāua shell lures may not have been strong enough to hold large oceanic pelagic fishes such as tuna (Scombridae) or kingfish (haku, *Seriola*

FIG.39 **Pā or trolling lure collected during the Cook voyages. Georg-August-Universität Göttingen, Oz 335**

lalandi), but they would have secured smaller coastal pelagic species including kahawai and koheru.[2] Using these lures, individual fishermen could take between 200 and 300 kahawai at a river mouth on each incoming tide.[8]

Early kahawai lures were made using a shank of bone, and were similar to, but shorter than wooden pohau mangā. Trolling lures collected by early explorers in New Zealand are limited to wooden barracouta lures (pohau mangā) and straight-shank bone lures (without pāua shell inlay), as well as a few lures made using the rim of a pāua shell for the shank. In contrast, archaeological examples in New Zealand museums are mostly incomplete stone minnow shanks (i.e. stone shanks lacking bone points), and later examples (post-1860) in museums throughout New Zealand and Europe are almost entirely wooden-backed pāua shell lures.

Post-European contact pā kahawai in museum collections are made with pāua shell lashed directly to wire, which formed both the hook point and the shank, and post-European wooden pohau mangā were made with a nail to replace the bone point.[258]

Ethnologist Roger Duff considered the wood-backed pā kahawai lure to be a relatively modern product because of its absence from the South Island.[259] Alternatively, it has been suggested that these lures are rare in the south because of the rarity of kahawai in southern areas.[223] Kahawai were previously common in southern waters and shoals consisting of millions were reported in waters off the Otago Peninsula with large numbers entering the harbours during summer months, but the species declined in abundance during the 1930s[156] and is now uncommon south of Banks Peninsula.[260] Recent anecdotal evidence suggests that kahawai stocks around the North Island have also collapsed following intensive commercial fishing,[261] although the kahawai stock in this region is at present considered to be at a sustainable level.[262]

South Island examples of pā kahawai constructed with bone or shell shanks are extremely rare in archaeological sites.[223] Wooden hooks or shanks rarely survive in early archaeological sites and frequently the only indication of the presence of pā kahawai or pohau mangā are bone points, which are common in sites throughout New Zealand.[223] Few wooden pā kahawai with pāua shell inlay in museum collections have provenance details, but those that do are primarily from the Taranaki region of the North Island, and most were made in the inland village of Parihaka where there was known manufacture of replica artefacts in association with European dealers in Māori artefacts in the late 1800s.[263]

Throughout tropical Polynesia, unbaited composite trolling lures made from shell, occasionally backed with whale bone, were used specifically to catch scombrids such as tuna (e.g. bonito, *Sarda chiliensis*), although it may have been a relatively recent development at some locations, such as Niué.[264, 265] The use of trolling lures may not have been extensive in some regions, and in some islands of Micronesia the traditional pearl shell lures possibly had a secondary function as a valuable or type of money, as almost all pearl shell lures have been recovered from archaeological sites with mortuary contexts.[191, 266]

Many 18th century examples of Polynesian trolling lures are represented in museum collections, and were made using pearl-oyster shell (*Pinctada* spp.) which was frequently exchanged over long distances.[267] These often very delicately carved hooks were of great value, particularly in French Polynesia, and were considered to be a special gift for a guest at the time of Cook's and other European voyages of exploration into the Pacific.[21, 187] Numerous examples of Polynesian tuna lures were gifted to the European explorers and are well represented in museum collections (Fig.40).[142]

Although made from brighter and more vibrant New Zealand pāua (abalone) shell, Māori shell lures (pā kahawai) do not appear to have been as highly regarded as suitable gifts, or were rare; examples obtained by pre-19th century explorers in collections cannot be verified.[142] James Cook collected trolling lures from a number of Polynesian islands, including Tahiti and Tonga, and some of these lures are now in the Institute of Cultural and Social Anthropology, University of Göttingen, Germany.[5] These are composite fishing lures with three parts: each has a shank of whale bone or wood; a 3-4 mm thick, whitish to reddish-brown glossy slice of mother-of-pearl from the tropical pearl oyster, attached to the outer (or convex) side of the lure; and points made of turtle shell, each with one barb on the innermost side.

The points have two holes at the base for attaching to the shank and trolling line with whitish fau-fibres (*Hibiscus tiliaceus*). A twisted cord runs from the upper hole at the base of the hook to a hole at the upper end of the shank, along the inner side of the fish-hook, and is further secured by an additional cord. From this second cord, a braided or simply twisted fishing line originates, thus the strain on the point from the hooked fish is taken up directly by the line itself, reducing the risk of the point slipping off the end of the lure.

FIG.40 **Polynesian tuna lure (Tonga): whale-bone shank backed externally with pearl-oyster shell, turtle-shell point. 153 x 58 mm. Date unknown. Museum of New Zealand Te Papa Tongarewa, Wellington, FE 007444**

Unlike mother-of-pearl shell, the iridescent colours of New Zealand pāua shell are on the concave, or inner side of the shell; hence pāua shell had to be used as an inlay rather than as a backing. The concave shape of pā kahawai prevented any supporting cord or snood between the barb and the fishing line, and firm lashings were required to retain the barb on the shank under the strain of a fish.

The earliest pāua shell trolling lures in collections (pre-1840) comprise hooks made with pāua shell rim shanks with simple non-barbed points often made from dogfish dorsal fin spines or

[5] Oz 211, Oz 213 and Oz 216

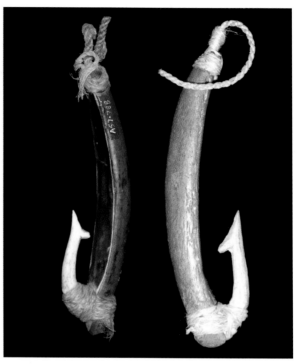

LEFT FIG.41 Pā kahawai with both shank and point made from pāua shell rim. Date unknown. Museum of Anthropology and Archaeology, Cambridge, 1899-2731 RIGHT FIG.42 Wood-backed pā kahawai with bone and wood points respectively. Dates unknown. Puke Ariki, New Plymouth: A, A57-788; B, A57-876

shell (Fig.41). Examples of lures made with straight whale-bone shanks and no pāua shell lining are represented in museum collections and were illustrated by Sydney Parkinson from Cook's first voyage (Parkinson 1773, plate XXVI, fig.4),[22] and Antoine Chazel from an example first brought back to France from Louis Isidore Duperry's visit to the Bay of Islands in 1824 on the corvette *Coquille*.[268] A few examples of curved whale-bone shanks with inlaid pāua are present in museum collections, but none of these whale-bone examples have provenance details or known dates of manufacture. It is probable that wooden or whale-bone pā inlaid with pāua shell were not easily manufactured until metal tools became readily available to fit the shell into the backing.

Trolling lures collected by Cook and other early explorers in New Zealand are limited to wooden barracouta lures (pohau mangā). No pā kahawai or pāua shell lures (either simple shell shank, or lures made from shell backed with wood) can be positively attributed to Cook. Given the value of attractive shell lures elsewhere in the Pacific, it is surprising that no examples of pāua shell pā kahawai were gifted to James Cook, collected during Cook's voyages, or obtained by other early explorers in New Zealand waters.

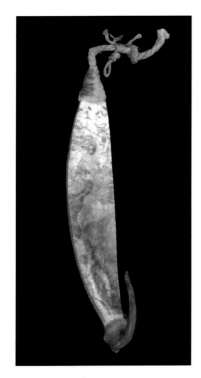

FIG.43 **Metal pā kahawai made using wire backing lashed to pāua shell. Date unknown. Museum of New Zealand Te Papa Tongarewa, Wellington, OL000106**

Wood-backed pāua shell lures (Fig.42) are unknown from archaeological sites in New Zealand; however, numerous incomplete stone minnow shank lures are well represented archaeologically. Examples of stone shank fishing lures complete with intact lashings and points from the Pacific region are rare: to date only one example from Tonga is known among the European collections,[6] and only one Māori example is known in New Zealand collections.[7]

Post-European examples of pā kahawai were usually made using wire (Fig.43), which enabled the wire to form both the shank and point by curving around the rear of the pāua shell which was securely lashed in place.[8, 56, 258]

Pā kahawai lures made with pāua shell backed with wooden shanks (and points with reverse barbs) were widely sought after by collectors in the late 1800s and are common in museum collections both in New Zealand[2] and Europe.[142] The earliest known example of a wood-backed pāua shell lure is held in the National Museum of Scotland[8] and was collected at the Bay of Islands during the *Rattlesnake* Expedition, 1846-50. The earliest known examples in New Zealand collections date from the late 1860s and were collected at Whanganui in the lower North Island.

Historical records clearly indicate that pāua shell was used in the manufacture of lures, but few of these records provide details of how the lures observed were constructed. Fish-shaped pieces of wood, inlaid with the shell of muttonfish (pāua) were described being used as lures by Māori at Petone, Wellington Harbour around 1839,[82, 84] while in 1843 Ernst Dieffenbach described a lure "with a navicular (canoe-shaped) piece of wood, lined on one side with a thin plate of pawa-shell (*Haliotis*) [sic], in imitation of a fish."[23]

George French Angas illustrated what appears to be a wooden lure with pāua shell inlay from his brief visit to New Zealand in 1844,[269] and earlier illustrations or references to wooden lures with pāua shell inlays (prior to the 1840s) are unknown. In 1807 John Savage referred to hooks

---

[6] National Museum of Ireland #1923 338b
[7] Auckland War Memorial Museum #5369
[8] National Museum of Scotland AO6256

made from the "…well polished outer rim of ear-shell…" (pāua),[60] but made no mention of wooden lures backed with pāua shell.

Metal pā kahawai are well represented in museum collections and the earliest examples date from the mid-19th century. The majority were made using iron nails, wire or copper that was beaten on an anvil or stone to flatten and curve the metal before a suitable section of pāua shell was lashed to it.[258]

Pohau mangā: barracouta lures

Pohau mangā were used to catch barracouta and comprised of a straight piece of reddish wood, usually beech (tawhai, *Nothofagus* spp.) or rimu ('red pine', *Dacrydium cupressinum*), with a simple non-barbed bone point embedded at the distal end (Fig.44). The wooden shank was generally plain, with a recessed groove at the proximal end for attaching the fishing line. The sharp teeth of the barracouta would easily cut the flax lines, so these lures were much longer than other lures and unlike lures in which the point of the hook is strengthened by attaching it directly to the fishing line. This was achieved by extending the line down the inner face of the shank, attaching the unbarbed bone point to the shank by inserting it through a hole in the shank and securing it with a wooden pin through a hole at the base of the point. A ridge in the bone point prevented it slipping through the hole in the shank. Pohau mangā (also called okooko)[37] were not decorated with pāua shell, but usually had tufts of feathers attached. As with other lures, the bone points of pohau mangā were rapidly replaced with metal nails after the arrival of Europeans (Fig.45).

Early examples of pā kahawai and pohau mangā

FIG.44 **Pohau mangā (barracouta lures) were made using long pieces of wood to protect the line from the sharp teeth of the fish. 180 x 30 mm. Date unknown. Museum of New Zealand Te Papa Tongarewa, Wellington, ME003974**

are notable in being made from very dense, hard wood, unlike later examples dating from the 1880s onwards which are generally made from a light wood, possibly tōtara (*Podocarpus totara*). Kahikatea (*Podocarpus elatus*) is one of the few New Zealand timbers that does not float when

LEFT FIG.45 **Bone points of pohau mangā were rapidly replaced with metal nails after European contact. Date unknown. Puke Ariki, New Plymouth, A47-157** RIGHT FIG.46 **Moa bone pā kahawai. 95 x 26 mm. Date unknown. Museum of New Zealand Te Papa Tongarewa, Wellington, ME 002227**

green,[270] although when dried it is extremely buoyant.[67] A lure which sank would be far more effective than lures that would float or skim along the surface of the water. Further study of the types of wood used may shed useful information on the manufacture of the original and later replica lures.

Several lures, each with a sharp barb set at an angle to the decorated shank, were trolled behind a canoe, which was then paddled rapidly through a school of fish. Lures made using stone or bone shanks would have fished deeper and been more effective than lures made using shell or wooden shanks, because of their weight. Four lures made using stained moa bone (without pāua shell inlays), and distinctive points from the Beasley collection in the British Museum are unique and this style of lure is represented by only two other examples, in the Museum of

New Zealand Te Papa Tongarewa collections (Fig.46). No similar lures are known in any other museum collection in Europe or New Zealand.

Another technique was to use the lure on the end of a short rod and line, without bait. This fishing method was known as kaihau mangā[1] and its aim was to snare the fish, then maintain pressure on the line so that the fish could not obtain any slack and disgorge the hook. As the fish struck the lure, the short rod enabled tension to be kept on the line and the fish was quickly flicked into the canoe; any relaxation of the line tension would enable it to escape. Two men using this technique, with one rowing and the other fishing, could catch 30 to 40 dozen barracouta within 2 or 3 hours.[88]

In 1827 a sealer, John Boultbee, reported rods used to take barracouta in Fouveaux Strait were up to 12 feet long, but with a short line of only 3 feet attached to the wooden lure. This particular technique relied on the long rods being used to move the lure through the water, while the canoe remained stationary but was equally as effective.[271]

Minnow shanks

Triangular, rounded or grooved stone or shell shanks are associated with early Polynesian-style hooks from elsewhere in the Pacific,[37] are known as minnow shanks (Fig.47) and are ascribed to the archaic period of Māori culture,[36, 251] while shanks made of bone and wood are more recent. [36, 223, 251, 272] Regional differences in the shape of stone shanks have been documented: those triangular in cross-section were more common in the South Island (particularly Marlborough), with examples also known from the Wellington and Taranaki regions, but are rare in the northern and eastern North Island, while rounded shanks were common throughout New Zealand, and grooved shanks were predominantly found in northern areas.[26, 36, 37, 140, 223, 231]

Early stone fish-hook shanks resembled those of hooks used elsewhere in Polynesia, more than the shanks of lures used by more modern Māori.[140] In 1908, Augustus Hamilton considered stone minnow lures to be charm stones, used without barbs to attract fish.[101] Others have suggested that it is hard to believe Māori would spend 100 hours carving a lure from stone and then risk losing it by trying to catch a mackerel.[44] Any lure, used with or without a barb, is at risk of being lost to sharp-toothed fish: an unbarbed 'charm-stone' is as likely to be lost as a fully barbed lure, no matter how many hours are spent in its manufacture. Lures (and lead sinkers) used on modern monofilament fishing lines are often lost unless a steel trace is also used.[2] It is unlikely that the stone minnow shanks would have had any magico-religious observance, and the sheer number of stone lures discovered, suggests they were actually used in fishing.

Few complete lures with stone shanks are known from the historical period in New Zealand,[37, 44, 101] and no archaeological examples exist. Kouaha, a poisonous gum from the bark of pukapuka, was used to preserve lashings on hooks (and later to prevent iron hooks from

FIG.47 **Stone minnow shanks. Dates unknown. Puke Ariki, New Plymouth, from left: 63-991, A63-996, A63-984, A63-987, A63-982**

rusting).[86] Gummy material on the lashings of some bone and wooden lures in the British Museum, England, and Hunterian Museum, Scotland, may have been applied by collectors to preserve the lashing,[249] and the extent of kouaha use by Māori is unclear.

It is probable that the muka flax lashings used to bind the barb to stone minnow lures would have required regular replacement with use. Practicality would suggest that these minnow lures were discarded and rapidly replaced as European metals became available, as has been documented for other hook types and stone tools.[37, 101, 140] As with wooden hafted toki adzes,[37] flax components, especially harakeke lashings of the discarded minnow shanks, would quickly decay.[1] The presence of bone barbs with two drilled holes at the base in archaeological sites is evidence of the use of minnow lures – the posterior hole was used to lash the barb firmly to the base of the stone shank, while the anterior hole served as an attachment to the trolling line (which was also lashed to the stone shank at the proximal end), thus ensuring that the barb did not slip off the shank when hauling in a fish.

Jigs

Unusual, slender lures made with an unbacked shank of pāua shell rim, and long, slender barbless points made from wood, shell or the dorsal-spines of spiny dogfish (koinga, pioke, *Squalus* spp.)

FIG.48 Pā kahawai or jig made using the rim of pāua shell shank with a wood point. 81 x 29 mm. Date unknown. Museum of New Zealand Te Papa Tongarewa, Wellington, OL000106/10

are present in museum collections (Fig.48) and were illustrated by Sydney Parkinson, artist on Cook's first voyage of discovery (e.g. Fig.38).[22] These lures appear unsubstantial and unlikely to be capable of holding a large fish without breaking, and superficially bear some resemblance to modern squid jigs: they may have been used at night to jig for squid (wheketere, *Nototodarus* spp.) in coastal waters.

Squid lack any internal skeleton; therefore, no body parts would be preserved in middens, apart from chitinous beaks, which could persist only in ideal situations such as in limestone areas, or middens rich in mollusc shell. There has been little evidence, from claimants to the Waitangi Tribunal, of the taking of squid in traditional times; it is thought that knowledge of squid and its taking may have been lost to present generations.[92, 93] Octopus, on the other hand, was eaten and there are historical records, although there is only occasional evidence of this in the Waitangi Tribunal claimants' evidence. As with squid, octopus may have been eaten only by an ariki in the eldest line of descent.[93] Elsdon Best noted that small octopuses were taken by hand and described how one of his worthy informants cooked and ate the "gruesome body", but he made no mention of the taking of squid.[1]

Although there is no archaeological evidence that Māori caught squid, the fishery would have been readily accessible. Europeans were unlikely to have accompanied Māori on nocturnal fishing expeditions to target squid, particularly in small canoes.[53] Joel Polack, a colonist who resided in the Bay of Islands from 1831-37, then later returned to Auckland where he lived from 1842-56, referred to the taking of cuttlefish.[67] The use of the term 'cuttlefish' is in reference to squid, as this was a general term in wide use in the 19th century and true cuttlefish (*Sepia* spp.) do not occur in New Zealand waters.[9] In 1908 Augustus Hamilton reported the taking of squid on coastal reefs, and commented that they were set aside as food only to be enjoyed by important chiefs.[101]

---

[9] Dr. Bruce Marshall, Malacologist, Museum of New Zealand Te Papa Tongarewa, *pers. comm.*, 2010

5

Kupenga: *nets*

> *"...after having a little laugh at our seine, a common king's seine, shewd us one of theirs which was five fathoms deep. Its length we could only guess, as it was not stretched out, but it could not from its bulk be less than four or five hundred fathoms."*

Joseph Banks 1769 (Beaglehole 1955: 444)

Early European explorers and settlers in New Zealand were impressed by the huge fishing nets used by Māori, frequently describing seines that measured "several thousand feet", "1000 yards or more" or "95 chains (2090 yards), well over a mile" in length. To set and retrieve these nets required teams of between 500 and 1000 or more people.[1, 17-20, 64, 67, 85, 273]

The large seines were used to encircle shoals of fish, which were then pulled ashore. In some instances when the catch was too large, the net was impossible to retrieve so the ends were pulled inshore as far as possible and tied to stakes until the receding tide made it possible to gather the catch.[1, 143]

Different nets were developed and used in a variety of ways. Set nets could be used passively to entangle the fish by the spines or gills,[44] but seines (kaharoa), bag-shaped (ahuriri), and funnel nets (matarau) were more common, as well as numerous varieties of frame nets (kape) and scoop-nets with pole handles (tikoko).[1, 35, 103]

Large funnel nets were employed particularly in tidal rivers where vast quantities of fish could be taken. Smaller funnel nets, baited and lifted vertically, were used from canoes. Using a funnel net from a canoe, an expert fisherman could take several hundred fish at once; in 1926 Sir Peter Buck noted that a haul of less than 700 blue maomao (*Scorpis violaceus*) was considered a poor catch.[103] The success of the customary netting techniques compared with those of European fishermen was often explained by the fact that Māori carefully watched the approach of shoals, instead of casting their nets haphazardly.[1, 143]

During his first voyage, James Cook described Māori fishing nets seen in the villages at the Bay of Islands:

"…we caught but few fish while we lay there, but procured great plenty from the natives, who were extremely expert in fishing, and displayed great ingenuity in the form of their nets, which were made of a kind of grass; they were two or three hundred fathoms in length, and remarkably strong, and they have them in such plenty that it is scarcely possible to go a hundred yards without meeting with numbers lying in heaps."[17]

Joseph Banks also noted fishing nets in the district north of the Hauraki Gulf:

"Fishing seems to be the chief business of life in this part of the country; we saw about all their towns a greater number of nets, laid in heaps like hay-cocks, and thatched over, and almost every house you go into has nets in process of making… Nets for fishing they make in the same manner as ours, of an amazing size; a seine seems to be the joint work of the whole town, and, I suppose, the joint property. Of these I think I have seen as large as ever I saw in Europe."[19]

In journals of the 1772 French exploratory voyages of the *Mascarin* and *Marquis de Castries*, both First-lieutenant Julian Crozet and Lieutenant Jean Roux described the structure of the Māori nets:

"…from 90 to 100 fathoms in length, and 5 to 6 in height. At the bottom is a case or basket in which are stones wherewith to sink the net… All along the top, at intervals, are little pieces of a round and very light wood, which take the place of the corks which we employ as floats… The knots of these seines are exactly similar to those of our nets…"[1, 66, 69]

Nets were generally made using unscraped flax, and the work frequently engaged all the inhabitants of a village.[1, 67] In 1885 Captain Gilbert Mair described a large net that was made at Maketu under the direction of chief Te Pokiha:

"Several hundred people were engaged in making the net which was constructed in numerous sections. Once completed, the sections were assembled and joined together in a process termed *toronga*. The upper and lower ropes were of tightly-twisted cabbage tree leaves which were stronger and more durable than flax. As these very large nets could only be carried by double canoes (*taurua*) two long single canoes were placed side by side and secured in position by means of poles lashed securely across them. A platform was then constructed across the width of the two canoes on which the net was piled..."

A crew of 30 men managed the vessel and net. An old expert or tohunga, positioned on a high vantage point onshore, gave the signal for fishing to begin. He allowed shoal after shoal of fish to pass, to the disappointment of many observers; when an apparently small shoal appeared he gave the word, "Haukotia mai" (intercept it).

"The *waka taurua* swung out across the front of the advancing shoal as the seine-tenders payed out the huge net, which, however, was not wholly expended. The spectators, not less than a thousand persons, were unable to haul the net. The spare ends of the seine had to be doubled back to reinforce the centre, and twice the net had to be lifted in order to allow a large part of the catch to escape. At full tide the great net was hauled in as far as possible and secured to stout posts until the receding tide left it and the multitudes of Tangaroa out of water..."[143]

Mair reported that some 37,000 fish were tallied, not including many "small-fry" and a number of sharks. The mesh of a large seine differed considerably in size in different sections of the net, being smaller toward the middle than at the ends, where it was required to be much stronger to support the weight of the catch.

Net-making

Details of net-making are provided by Sir Peter Buck in a 1926 report published in the *Transactions of the Royal Society of New Zealand*.[103] The information was gathered from Ngāti Porou of the east coast of the North Island from north of Gisborne to Lotin Point during a Dominion Museum ethnological expedition to the district in 1923. The expedition also included Elsdon Best, who reported how the nets were used and associated ceremonies.[1]

Much of the information was gathered at Waiomatatini and Rangitukia, in the Waiapu Valley, and at Te Araroa, on the coast. Additional notes were obtained from Whanau-Apanui in the neighbouring stretch of coast in the Bay of Plenty.[103] A wide range of nets were used in different regions of New Zealand and terminology varied; hence, names for nets provided by Buck and Best may differ from those used in other areas.

Despite the profusion of traditional flax nets that were used until well into the 20th century, only a few (mostly smaller hand-nets), have been preserved. Museum collections consist mainly of fragments found in dry archaeological sites or waterlogged swamps.[36] Consequently, details of knots and net-making techniques in different regions of New Zealand are poorly understood, although some information has been gathered from comparisons with similar Polynesian nets.[44, 274]

Large nets were made from leaves of undressed (green) flax which were split into narrow strips, semi-dried, then knotted together as required. Smaller nets were made of twine prepared from dressed muka fibre. Muka twine could be rolled into a ball and was passed through each mesh as the rows were made. A mesh-gauge (papa kupenga, or karau), fashioned from wood, or from a piece of whale bone, was sometimes employed by the weaver (kaita), but skilled net-makers used their fingers as a gauge.

Floats for the top rope of the net were usually fashioned from some light wood, the most suitable being whau, or houama, a small tree (*Entelea arborescens*). Occasionally, floats were made from dried and rolled up raupo leaves (although these were not durable),[24] or pumice stones. Sinkers for the lower, or foot rope, were long, smooth, water-worn stones of a suitable size enclosed in a netted sheath or bag (kopua), which was attached to the lower edge of the net.[1, 66, 103]

Elsdon Best described the process of net-making:

"When about to commence the weaving of a section of a net, a strong plaited cord, (*ngakau*), is doubled and secured by one end to a peg at such height as is convenient to the net-maker, who sits down to his task. On this looped cord the first line of meshes is made. As the netter proceeds with his work he does not leave the meshes properly spaced, or he would be compelled to frequently change his position; he pushes them to the left along the *ngakau* cord so that they become bunched together near the peg. In forming a mesh the operator holds the gauge in position, passes the strip of material over it, and hitches it on to the lower part of the adjacent mesh in the upper row. When the section is completed, then the *ngakau* cord is withdrawn. The first row of meshes made is called the *whakamata*. The term *torea* is applied to an extra mesh formed in order to increase the width of a netted fabric, as in the making of a funnel-shaped net.

In some districts, when an important new net was to be made, a party of men would proceed to the flax-grove in order to procure material. Each man would be clad only in a form of kilt, and these would be new garments made for the occasion. These men would each bring to the village a large bundle of material, leaves of *Phormium* divided into narrow strips, and these would be hung up until all was ready. On the following day a party of women would proceed to the place and procure another supply. On the third day the net-making would commence. When, in the process of manufacture, a netter had to attach another length of flax-strip he did not cut off the protruding ends of the splice. If in making the knot his finger became caught in the loop, it was a sign that when the net came to be used a fish would be caught by the head in that part of the net. But others say that it was an evil omen: that the man whose finger was so caught would die ere the net was used. Such beliefs, omens, &c., differ considerably in different districts, and, indeed, one may hear differing explanations in one village.

Some speak of the strips of flax employed in net-making as having been smoked in order to make them more durable. Captain Johnstone mentions a needle, but the Māori maker of green-flax nets used no needle at his task. When watching a Whanganui native making a net I noted that he placed one finger of the left hand through the mesh above and pressed downward so as to pull it taut. He then with his right hand formed the new mesh over the next finger, extending in a similar manner and spaced by eye; thus in this case the size of the mesh was regulated by the eye alone, the result being satisfactory, so far as a tyro may judge.

When the various sections of a new net were assembled the long upper and lower ropes were stretched out and attached, and the floats and weights were secured. The net was not folded up for carrying to the vessel on which it was to be stowed, but was carried loose and extended. The net lay fully extended when the assembling was completed. A row of bearers took station on the western side of the long net, the bearers being about 2 fathoms apart. The expert took his stand on the eastern side, and when he gave the command *"Hapainga!"* each man grasped the net (with his left hand first) and swung it up on to his left shoulder. The expert now took his place at the head of the procession and gave the word to march. On reaching the canoe two men took up the task of stowing the net on board; it was so stowed that the upper rope of the net was on the right-hand side of the vessel. As each man handed over his part of the net for stowing he turned to the right and stepped aside. It would be deemed an unlucky act were he to turn round to the left.

After nets had been used they were dried by being hung over a long railing. When stowed away they were folded up on elevated platforms and a roof put over them in order to protect them from the weather…"[1, 275]

Types of nets

Large seines were most frequently commented on by early Europeans; however, many smaller nets were made and used, including drag-nets, scoop-nets, and other designs. In tidal rivers two principal types of net were employed: drag-nets (kaharoa) stretched across from bank to bank, and large funnel-shaped nets (ahuriri) which were set horizontally in the water channel and held open by the flow of water.

The funnel nets were made in a variety of forms and could be up to 22m (75 ft) in length and

FIG.49 **A group of Māori men holding up a large matarau fishing net. Photograph taken between 1900s and 1920s, by J. McDonald. Alexander Turnbull Library, Wellington, New Zealand, PAColl-8238**

7.5m (25 ft) in diameter at the mouth. The funnel narrowed down to 45 cm (18 inches) before leading into a huge basket capable of holding up to "…two hogheads…" (500 litres) of fish. These nets were set near the mouth of a creek in the tidestream, and held in position by two stout spars firmly driven into the river-bed, and were filled to tightness each tide…"[1, 275]

At sea, smaller funnel nets (matarau and tarawa) held open by means of hoops were fished by lifting the net vertically from the seabed to the surface and could be used around rocky coastlines unsuitable for drag-nets (Fig.49).[1, 103, 275] The net was raised by a strong rope attached to two cords that were secured at four points to the hoop that held the entrance of the net open. Bait was suspended by short strings (tau mounu), or tied to the bottom of the net, and a stone sinker was placed at the bottom of the net. James Cook described how the funnel net was used in the South Island at Queen Charlotte Sound during his first voyage in 1770:

"As we were returning we saw a single man in a canoe fishing; we rowed up to him, and, to our great surprise, he took not the least notice of us, but even when we were alongside of him continued to follow his occupation without adverting to us any more than if we had been invisible… We requested him to draw up his net, that we might examine it… It was of a circular form, extended by two hoops, and about seven or eight feet in diameter; the top was open, and sea-ears [pāua] were fastened to the bottom as a bait. This he let down so as to lie upon the ground, and when he thought fish enough were assembled over it he drew it up by a very gentle and even motion, so that the fish rose with it, scarcely sensible that they were being lifted, till they came very near the surface of the water, and then were brought out in the net by a sudden jerk…"[17]

This was probably the same net described by Joseph Banks:

"It is circular, seven or eight feet in diameter, and two or three deep; it is stretched by two or three hoops, and open at the top for nearly, but not quite, its whole extent. On the bottom is fastened the bait, a little basket containing the guts, &c., of fish, and sea-ears, which are tied to different parts of the net. This is let down to the bottom where the fish are, and when enough are supposed to be gathered together it is drawn up with a very gentle motion, by which means the fish are insensibly lifted from the bottom. In this manner I have seen them take vast numbers of fish, and indeed it is a most general way of fishing all over the coast…"[17, 19, 273]

Elsdon Best described details of a matarau funnel net made at Waiapu:

"Pliable stems of climbing-plants were used to form the hoops employed in distending this form of net. Across the wide mouth of the net were secured certain cords, and to the middle parts of these were attached the stout cord by means of which the net was lowered and raised. The bait was secured to the cross-cords, and a stone sinker was placed at the bottom of the net, while others were secured to the hoops. The net was made of twine composed of dressed flax-fibre but was reinforced with strips of undressed flax-leaf. A later note from the same informant states that the upper part of the net was of fibre, and the lower part of raw material, of which the upper part was the most durable. Such a fish as the *ururoa* [sharks] might break through the lower part of the net, but not through the upper part. This informant also remarked that when the net was hauled up it was done in a vigorous manner, the result being that the fish were forced to the bottom of the net – forced downwards by the resistance of the water as the net was pulled quickly up. When the hoop appeared above water it was so manipulated as to turn round several times. This twisted the net-fabric and so brought its lower part, containing the fish, nearer to the surface. A length of rope secured to the bottom of the net was then brought to the surface by means of a hooked rod, and by pulling on this rope the bottom of the net was hauled up to the surface and dragged into the canoe; then the lower end of the net was opened and the fish emptied into the hold…"[1, 35]

Small hand-nets were used extensively in shallow reefs and intertidal areas. Toemi was a small circular bag-net held open by cross-sticks arranged like the spokes of a wheel, while the mesh projected above the top hoop. The free edges of the mesh were drawn together by the same cord that was used to pull the net up – as the fisher raised the net, the cord closed the mouth of the net, preventing the fish escaping.

Scoop-nets (kape) were used for taking fish in shallow waters. In northern areas large scoop-nets were used in the surf to catch mullet and kahawai (Fig.50). Both Elsdon Best and Sir Peter Buck described the taking of marblefish on the east coast of the North Island with small hand-nets on poles. Several fishermen with nets would station themselves at the end of channels and block any escape routes. Divers or others armed with long poles would drive out fish from beneath the rocks and into the nets.[1, 103]

FIG.50 **Large scoop-net for taking kahawai and mullet. Museum of New Zealand Te Papa Tongarewa, Wellington, negative A4070 by J. McDonald**

Hīnaki: pots

A wide variety of pots and traps were used by Māori to take crayfish on coastal reefs. Pots were the most important method of taking eels and lamprey in freshwater, and they were also used to take marine fish (Fig.51, 52). Hīnaki were fished as baited traps, or could be placed in natural rock channels, or in gaps in artificial weirs where fish were swept into them by the force of the water.[167, 170] Methods of construction and how hīnaki were used have been described in detail by Sir Peter Buck[103] and Elsdon Best.[1]

Hīnaki were constructed using a wide variety of suitable materials. Kiekie was the most common material used, and was the easiest to obtain. However, it was also the least durable – even with care, pots lasted only five to seven years. Other climbing vines including white rata vine (akatea, *Metrosideros albiflora*), wire grass (tororaro, *Muehlenbeckia astonii*) or climbing fern were sought after as they gave the best results, both in strength and durability.

FIG.51 Traditional Māori wicker baskets for trapping eels and other fish. Head, Samuel Heath, d 1948 :Negatives. Alexander Turnbull Library, Wellington, New Zealand, Ref: 1/1-007414-G

Aerial roots of kiekie and selected vines steeped in water until pliable, were light, strong and flexible, and could be woven into fine, flexible, springy baskets. Bindweed (pohue *Calystegia sepium*) was also used in the construction of hīnaki, but was limited in strength and durability. Supplejack (kareao, *Ripogonum scandens*) and slender rods from tea tree (mānuka, *Leptospermum scoparium*) were preferred for making frames of crayfish pots as the stout canes could withstand buffeting by waves in the coastal reefs.[103, 276]

Many of the older hīnaki in museum collections are black in colour. This is a result of a tanning process that was used to extend the life of the trap. Quantities of bark from black maire (*Nestegis cunninghamii*) and hīnau (*Elaeocarpus dentatus*) were gathered then wrapped in leaves, and steamed in an earth oven (umu). A special trough, called patua, made of inner bark taken from a large totara tree, was prepared (a bark trough would not

FIG.52 **A wicker fishing pot (hīnaki).** McDonald, James Ingram, 1865-1935: Photographs. Alexander Turnbull Library, Wellington, New Zealand, PA1-q-257-71-1

absorb the tannining solution as a trough cut out of wood). The bark was first cut to the required length with an adze, then peeled off in one sheet without splitting, using a wooden lever with a fire-hardened point made especially for the purpose. The ends of the sheet of bark were gradually softened by steam in an oven until they were quite pliable and could be bunched and tied to form a trough. The softened pieces of prepared hīnau and black maire bark were crumbled and broken as small as possible, then covered with water and macerated in the patua to extract the tannin. Bundles of vines intended for making hīnaki were soaked in the resulting solution for one or two nights, according to the thickness of the vines.[87]

Although wire mesh soon replaced traditional fibres for weaving materials after the arrival of Europeans, traditional materials were used well into the 20th century and today large cray-pots are still made using supplejack hoops, held in shape with wire.

FIG.53 A reconstruction illustrating the legend of the creation of New Zealand, showing the demi-god Māui fishing the North Island up out of the sea. Photolithograph by William Dittmer, 1907. Alexander Turnbull Library, Wellington, New Zealand, PUBL-0088-049

6

Ritenga: *ceremonies and fishing activities*

"Too little has been said and too little is known of the way in which stone implements were made and used... When the savage acquires an axe of steel his beautiful but ineffective stone weapon becomes useless, and falls from his hand... and by the time a man who not only feels a little curiosity on the subject but desires to impart a little information to his curious countrymen dwelling in the remote Old World comes round, the savage and his savage children have gone to shadow-land..."

Merchant and bookseller George Chapman, 1881 (Chapman 1881: 481)

In Māori mythology, Tangaroa, a child of Ranginui (sky father) and Papatuanuku (earth mother) was regarded as the patron of all fish, to whom fishermen made offerings and repeated charms to ensure good luck, and the legend of the creation of New Zealand depicts the North Island as a fish, hauled up by the demi-god Māui (Fig.53).[167]

Throughout New Zealand, both fishing activity and the manufacture of fishing equipment were regarded as strictly tapu to ensure the gods were placated, and the fishing would be successful. Karakia were recited to attract fish to the fishing grounds and to ensure hooks, lines and nets operated effectively. Karakia were also recited over canoes, particularly in connection with sea-going voyages, to ensure good passage and safe arrival at the destination. In the event of a storm arising a tohunga (expert or priest) could recite a karakia to secure fine weather or reduce the strength of the wind.[1, 8]

Ceremonies for opening the fishing season as observed by Māori in the North Island, particularly in the Bay of Plenty, Poverty Bay, and Taranaki regions, have been described and recorded by early historians.[1, 56, 249] Each year, on the day appointed for the opening of the season, a canoe set aside for the purpose was prepared with required ceremonies and incantations, and the tohunga then placed a punga-tai in the canoe. The punga-tai was a small cup or basin-shaped object generally made from a block of pumice or a light porous stone, and filled with earth from an umu or earth oven, in which the first fish of a previous fishing season had been cooked. The tohunga then set out for the fishing ground and before casting his line recited prayers over the fish-hooks and punga-tai. No other line was permitted to be used until the first fish had been caught. This first fish was sometimes released, or if it was a desirable food species it was retained and cooked in a special oven and eaten as a sacrificial feast to Tangaroa by high-ranking tohunga and other chiefs of the tribe.

Because of the dangers and risks associated with sea fishing, tapu prohibitions were particularly strict and men proceeding out to sea in fishing canoes were not permitted to take any food with them. Generally, sea fishes were represented by a mauri or talisman, which was usually a stone concealed on some part of the coast belonging to the iwi. As a kind of shrine or emblem of the gods it represented their powers and this attracted and conserved fish. Fishing canoes also had their own talisman, which retained the tapu of the vessel, and protected it at sea. If the canoe was repeatedly unlucky in fishing, then it was known that there was something wrong with the talisman and its tapu had been compromised.[1]

The customs and rituals varied from region to region and iwi to iwi. In the Bay of Plenty it was considered unlucky if a fish touched the gunwale of the canoe when it was being landed – the fish was then laid lengthwise in the canoe, and if any person stepped over it, some dire misfortune would result. When landing groper into a canoe, great care had to be taken to ensure the fish did not come into contact with the stone sinker, or any other stone in the canoe or no more fish would be caught that day.[8] When fishing for hāpuka it was regarded as unlucky to refer to the fish by name and it was alluded to indirectly as "rarawai".[123]

In northern Taranaki the mauri of the kahawai in the Waiongana district consisted of a small quantity of sea sand that had been rendered tapu by a ritual performed over it by a tohunga. The sand was kept in a stone cup (punga-tai), and a little sprinkled in the water by the fishermen attracted the kahawai to the canoe.[53]

In many regions, moki were regarded as a special or sacred fish, and strict rituals for catching this fish were observed. Preparations for the first day of fishing for moki in the Bay of Plenty were described by Tāmati Poata in 1919:

"…before proceeding to the fishing ground, the fisherman must dive for his crayfish bait, which must be left in the boat, together with the lines, after the boat has been thoroughly cleansed. If he takes his bait home and brings it down in the morning, he will not catch a single moki. He may catch other varieties, but the moki will refuse to bite. He must take neither food nor water with him, nor must he even think about food or mention it to his mates. His wife must not get up in the morning to light the fire for cooking purposes.

The mother must not under any circumstances suckle the baby at the breast. These rules apply to the opening day of the season, and if they are strictly carried out the fisherman may depend on a good supply right through…"[8, 56]

Ceremonies and net-making

According to mythology, there was a time when Māori were not acquainted with the art of net-making.[1]

"In olden times it was that one Kahukura acquired the art from a strange people, the Turehu folk. He chanced to come upon them as they were hauling a net under cover of night, and, owing to the darkness, he managed to join them and take a part in their task. His aim was to so delay them that they would be overtaken by daylight ere their task was completed; by these means he hoped to secure their net or a knowledge of its construction. The Turehu folk performed all tasks during the night, for in olden times the activities of many abnormal beings and supernormal objects had to cease when darkness passed. Kahukura effected his object by his deceitful manner of stringing the fish. He did not secure the cord so that it would retain the fish; hence, having strung a number, when he lifted the string all the fish slipped off it. He repeated this deceitful performance several times and thereby gained his object. He so delayed operations that ere the task of stringing the fish was completed day dawned, whereupon the Turehu folk fled in haste to their home among far forest-ranges, abandoning fish and net. This is how the Maori folk acquired the art of net-making; they had but to observe the fabric left by the Turehu folk in order to acquire that art. This episode is said to have occurred at a place called Rangiawhia, but it is probably an old-time story brought from other lands. The natives of Niue Island tell a similar story to account for their knowledge of the art of net-making. Our Maori folk have preserved the following saying referring to the deceitful act of Kahukura in stringing the fish: *Ko te tui whakapahuhu a Kahukura.*"[1] [The string united by Kahukura]

The manufacture of nets, particularly large kaharoa seines, was a task subject to extreme tapu[1, 103] and restrictions were enforced rigorously with offenders even slain, and canoes inadvertently passing an adjacent stream or tapu beach area destroyed. No fire or food preparation was permitted within the net-making area until the net had been completed and the restrictions lifted.

European observers suggested that this was to induce the workers to stick to their occupation and so remove the irksome restrictions as soon as possible;[67] however, Elsdon Best noted that such communal tasks were carried out willingly, and that the tapu restrictions arose from the cultural expectations of Māori in which the spiritual presence of gods (atua) was necessary, without which the undertaking could not succeed. Nets were often made for specific purposes, such as taking fish for particular guests and hence could not be used inappropriately.[1]

When a new net was about to be made, a tohunga would proceed to the place where the flax was being gathered, and he would pull two of the young central leaves of a flax plant until they broke off at the base. As he pulled out each leaf he repeated the words *"Tangaroa whitia, Hui – e! Taiki – e!"* The first blade represented the men of the community, and the second represented the women. If a screeching sound was heard as the leaves were torn out the omen was a good one – the net would be efficient. If no such sound was heard, then the net would not be a lucky one. In some districts the first day's work in cutting and preparing material was done by the men, and the second day's work by the women. If the butts of the plucked leaves were jagged it was said that the spirits of the fish yet to be caught in the net had nibbled them.[1]

When completed, the new net was inspected by the tohunga, who then cut off the loose, protruding ends of all splices where knots had been made in adding lengths of material, which the net-makers had not been permitted to cut off. Two ropes were then made; these were not used for the net, but were deposited by the tohunga, along with the severed ends from the knots, to the ahu or tapu place, while reciting a charm to bring good luck to the fishermen.

Two more ropes were then made and attached to the net before the tapu was lifted with befitting ceremony and karakia (following differing procedures in various districts). When the net was first used, the tohunga seized one of the fish in the net with his left hand, and, holding it with its head underwater, said *"Haere mai, haere ki tai nui no Whiro ki te whakataka mai i to tini, i to mano."* He then liberated the fish outside the net. This karakia asked the fish to go out to the great ocean, assemble its kind, and bring them to the net. In other rituals observed, the first fish landed could be burnt, or deposited in a sacred place as an offering to the gods, and then preparation of the first catch or catches was carried out for ceremonial feasts involving those who had worked on the net. During these ceremonies fish were cooked in steam-ovens for priests, experts, and those of chiefly rank, and in separate ovens for men and for women,[1] after which the net was used in the usual way.

Matau, aho: hooks, lines

Fishing rituals were closely associated with the hooks and lines that were used in fishing activity, and included many ceremonies associated with the first use of fishing tackle.[1, 35, 89, 140] Many matau have small detailed carvings on the outer loop where bait was attached, or at the snood end of the shank where the hook was lashed to the line (Fig.54); these have been interpreted as symbolising the god of the sea Tangaroa,[101, 140] despite at least some of the grooves interpreted as mouth-lines actually being grooves for fastening the fishing line to the hook.[101]

In the late 19th and early 20th centuries William Colenso, Elsdon Best and others attempted to document karakia and other oral traditions that had been passed down through generations of Māori.[1, 25, 26, 35, 51, 53, 89] By this time the impact of European culture had led to significant changes in the Māori lifestyle. Spiritual ceremonies were no longer observed regularly and the karakia and much of the knowledge held by elders, tohunga and chiefs had already been lost.

Māori ceremonies conducted before fishing expeditions involved the lines and hooks that were to be used to catch fish. When the fishing grounds had been reached, all the hooks were carefully arranged by being stuck in the raupo that covered the joints of the side plank of the canoe. The tohunga then recited another karakia to ensure that the fishing equipment functioned effectively. The first fish caught was returned to the sea, with a karakia uttered over it before release, to encourage it to bring an abundance of fish to the fishers' hooks. If kahawai were the only species taken, women were not allowed to partake in the meal; if snapper or other fish were caught, then the women were permitted to eat, but the first fish taken belonged to the tohunga.

FIG.54 **One-piece bone rotating hook with carved ornamentation on the head of the shank and bait string attachment point. 80 x 53 mm. Date unknown. Museum of New Zealand Te Papa Tongarewa, Wellington, ME002237**

On returning to shore, if the fishers had been very successful, three ovens were set aside under tapu: the first was called He Marae, for the elders; the second was called Te Ikahoka, for the priest of the canoe; the third was called Te Tukunga, for the remainder of the party. The tohunga took a fish and pulled out a piha (gill), then taking it to a sacred place, and holding it suspended by a string, he uttered another karakia to the gods.[89]

Elsdon Best described how all fishermen using lines were acquainted with at least one karakia for recital when fishing and these were repeated over fish-hooks; he documented an example[1] (see Chapter 3), while the Reverend Richard Taylor noted several religious ceremonies connected with fishing and described how, the day before Māori went to sea, they arranged all their hooks around some human excrement, and used a karakia "…which will not bear being repeated…"[89]

Elsdon Best also described the ceremony surrounding the use of a new fishing line:

"…when a man used a new fishing-line (aho, nape) for the first time in fishing (hi = to fish; usually as hi ika, the latter word meaning "fish") he went through a strange performance. Amid the silence of his brother craftsmen he tied a sinker (mahe) on his new line, and then the hooks, beginning with the lower one. He then baited the hooks, not forgetting to expectorate on each bait as he tied it. He then coiled up the line and passed it under his left thigh, after which the line was passed over the left side of the canoe in its first wetting. When the line was out he lifted it a little if the sinker had touched the bottom, held it in his left hand, and with his right hand, dipped up a little water and threw it against the line. When he caught his first fish he deposited it in the stern of the canoe, after which his companions were allowed to commence fishing. When the party returned to land, the owner of the new line took his first-caught fish and the fern or bulrush leaves he had used as a seat, and returned home. There he generated a fire by friction and burned the fern, and at that fire he roasted a portion of the gills of the fish taken from the right side. He then took the gills in his left hand, lifted it up and waved it to and fro, at the same time calling to his dead male relatives that here was food for them: it was an offering to the spirits of those defunct relatives. He did the same with the portion of gills from the left side of the fish: this was an offering to the spirits of his deceased female relatives. The fish he deposited at the tuahu…"[35]

Ornate and ceremonial hooks

Wooden composite hooks and bone hooks collected during the expeditions of James Cook and other European explorers prior to 1850, although skilfully made, are generally plain and robust, without carved adornment. Sydney Parkinson, artist on Cook's first voyage to New Zealand, illustrated nine hooks; only one has carving on the shank – an unusual but distinctive carved head, possibly a manaia (Fig.55).

One hook collected on Cook's second or third voyage has an ornate carved figure on the shank, and a second hook, also collected on the second or third voyage, has a carved mask on the snood knob (both held in the National Museum of Ireland). All other fish-hooks known to have been collected on the Cook voyages or by other 18th century European explorers are plain, without carved adornment. It is possible that other pre-contact hooks with ornate carving were

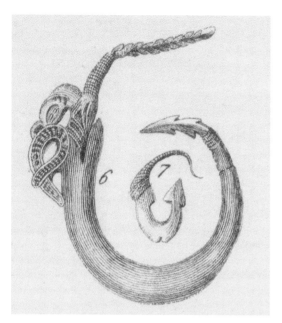

LEFT FIG.55 **Carved composite hook, and one-piece bone hook illustrated by Sydney Parkinson on Cook's first voyage. Detail from Parkinson 1784, PLXXVI** RIGHT FIG.56 **Ornately carved hook. Date unknown. © Trustees of the British Museum. British Museum, London, 1944Oc2.171**

collected by early European explorers, but remain in private collections and have not been passed on to public museums.[229]

Hooks obtained by museums later in the 19th century, particularly in the period from 1880-1910, show an increasing level of adornment, from simple carved snood knobs (koreke), to hooks with elaborate, often highly detailed carving on the shanks. Detailed ornamental carving of fish-hooks was not easily produced until steel tools became available after the arrival of Europeans, and the production of many ornate hooks in the late 1800s and early 1900s, by both Māori and European forgers, was in response to demand created by European dealers and collectors (Fig.56).[141, 142, 263]

Many ethnological fish-hooks in museum collections that appear to have been produced for artefact or curio trading, may be copies of earlier designs. The carving on some of these hooks is orientated laterally, rather than aligned with the point, suggesting the hooks may have been made for display purposes (Fig.57). The demarcation between hooks produced for fishing, trade with collectors, or personal decoration (hei matau), is difficult to determine, in part because of the unusual hook design, which is related to the rotating manner in which the circle-hook functioned.[2]

It has been suggested that some hooks were manufactured for use as ornaments or charms for ceremonial use,[101, 140, 277, 278] including use as magico-religious objects[139] and were not intended for use in fishing. Elsdon Best,[1, 167] Sir Peter Buck,[37] William Colenso,[24, 25] John White,[123] and others,[89] who documented known Māori mythology and rituals associated with fishing, made no mention of the manufacture or use of symbolic or ceremonial hooks. The English collector Harry Beasley described several unusual slender hooks that he considered to be for ceremonial purposes,[140] but these hooks were made to catch seabirds such as albatross and were perhaps restricted to the East Cape region.[1, 26, 101] The circle-hook design used by Māori, particularly when hooks were manufactured entirely from bone, or stone such as pounamu, has made interpretation of their use difficult.[2]

There are no pre-contact traditional Māori fish-hooks which can be interpreted as purely ceremonial in function in any museum collection. There is no evidence that Māori produced ornate hooks for ritual purposes. Māori ceremonies conducted before fishing expeditions involved the lines and hooks that were to be used to catch fish[142, 229] and all evidence suggests that hooks used in fishing were practical, without ornate decoration.

Hooks which have been described as ceremonial are likely to be late 19th century examples made for trade purposes. For example, two composite Māori fish-hooks are held in the Blackburn collection in Hawai'i.[278, 279] One hook is plain, lacking any carving, and has a reputed provenance to Joseph Banks on Cook's first voyage. The second hook has a richly carved mask, with pāua shell inlays, extending over almost half of the shank and has been described as a "ritual fish-hook". The provenance for this "ritual" hook was attributed to the London Missionary Society, based on an illustration in *The Story of the South Seas* by George Cousins in 1894;[280] however, it is not the same hook – the London Missionary Society hook is quite different in the lashing and it has a much smaller, crudely carved mask extending over only one fourth of the shank. The carved Blackburn hook – as well as the hook illustrated by the Missionary Society (Fig.58) – although probably authentic (in being made by Māori), are typical examples of hooks produced in the late 1800s for trading with tourists and collectors.

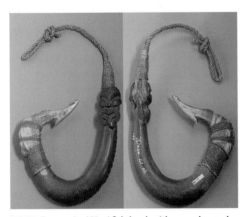

FIG.57 **Composite Māori fish-hook with carved snood knob. The lateral orientation of the carving suggests the hook may have been made for display purposes rather than fishing. 127 x 100 mm. Date unknown. Museum of New Zealand Te Papa Tongarewa, Wellington, OL000105**

One example of a pā kahawai in collections held at the Museum of New Zealand Te Papa Tongarewa (Fig.59) has been described (initially by the vendor) as a "ceremonial hook… made of the finest materials including sperm whale bone lined with pāua, and a greenstone point…"[277] However, the hook

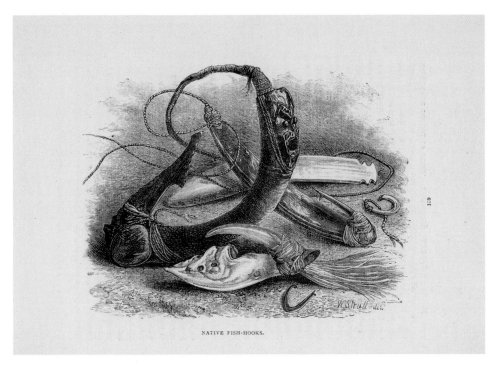

FIG.58 **Māori and Polynesian fish-hooks illustrated by the London Missionary Society. After Cousins 1894.**

has been made from a pig's tusk, not sperm whale tooth or bone, and the ornate greenstone point (showing signs of metal file marks) is a copy of a spear tip; it is unclear how the sections of pāua shell are held in place, and the lashings are carefully made of 3-ply plaited fibre (perhaps kiekie; or pīngao, golden sand sedge *Ficinia spiralis*), not traditional twisted muka cord. Although described as being made between 1800 and 1900, the hook has no known provenance beyond the collector. The components used and method of construction suggest that the hook is a forgery, and it may have been made as late as the early or even mid-20th century.

The hook was purchased for $40 by the museum in 1968 from a private vendor, Paul Webster, who had received the hook from London-based dealer and collector, Kenneth Athol Webster, who died in 1967. The collection history of Kenneth Webster is somewhat complex. Webster was a New Zealander who moved to London in 1936 and began collecting and dealing in Māori artefacts after the Second World War. In 1945 he received a letter of support from W.R.B. Oliver (dated 3 December 1945), Director of the Dominion Museum (now Museum of New Zealand Te Papa Tongarewa), that reads: "…The bearer, Mr Kenneth A. Webster, is forming, at his own expense, a collection of Māori artefacts. These he has undertaken eventually to present to the Dominion Museum. Accordingly, I have been authorized by the Management Committee of the Museum to state that it approves of the arrangement and would appreciate any assistance that could be given to Mr Webster. [signature] (W.R.B. Oliver) Director."[10]

In the post-war era of the 1940s and 50s, many museums in Britain lacked adequate funding, and collecting policies were being reviewed. Items that were no longer considered appropriate to collect and store were disposed of, often being dumped, sold at public auction to raise funds, or even donated to local dramatic societies for use as stage props. Webster became a self-appointed procurer of Māori artefacts for New Zealand museums, and collected items by purchase or exchange from the small English county and private museums.[281] Initially he proposed to gift or bequeath the entire collection of artefacts thus obtained to the Dominion Museum, and on several occasions wrote to the museum expressing this desire as space in his London home was limited.

Following concerns over items that he retained for his personal collection and sold to cover expenses, the letter of introduction was withdrawn by Robert Falla, who succeeded Oliver as Director of the Dominion Museum in 1947.

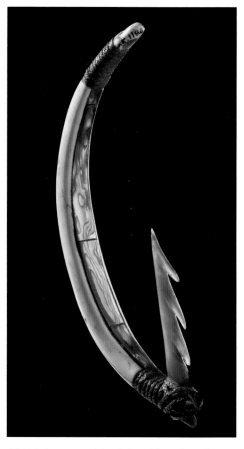

FIG.59 **Fake ceremonial pā kahawai from the Webster Collection. 153 x 53 mm. Date unknown. Museum of New Zealand Te Papa Tongarewa, Wellington, ME011848**

Webster continued to collect Māori taonga, and in the late 1940s sold several items to the museum as well as sending other items on loan for various periods of time, on occasion requesting their return for sale to other dealers and collectors. In the early 1950s, Webster approached the Leeds City Museum, again purporting to be collecting Māori material on behalf of the New Zealand Government, and was reportedly given a free run of the collections. He took not only Māori items, but virtually all the Pacific material and some of the best American and African material. He kept much of the material and sold the rest on the private market.[282, 283]

In 1954 and 1958 Webster deposited a collection of several hundred Māori artefacts on

---

[10] Museum of New Zealand Te Papa Tongarewa archives MU00412/001/008

permanent loan with the Dominion Museum, and this collection was subsequently bequeathed to the museum on his death in 1967,[283] although the government had to pay considerable British estate duties of over $12,500. In 1962 Webster had offered to sell the last part of his collection to the New Zealand Government but the offer was ignored by Falla, following misunderstandings in relation to some tattooed dried heads (mokamōkai) Webster had included in an earlier loan to the museum. The remainder of the collection was subsequently sold to a New Zealander, Mr Andrew Meit. After the death of this buyer, the collection was sold at auction in three parts in the early 21st century. Approximately 300 Māori items from the collection were sold in 2002,[284] and a further 482 items were sold at auctions in 2010 and 2011, returning almost $3 million in sales.[285]

It is unclear when or how Webster obtained the "ceremonial" pā kahawai or why it was not included among the items deposited at the museum in the 1950s. It is not included in an undated list held in the museum archives of approximately 500 Webster Māori Collection items. This list does include many items which are not identified but are marked "not packed". Webster is known to have met convicted forger James Edward Little after the latter's release from prison in 1947, although by this time Little had supposedly ended his career of producing forgeries,[286] and it is unlikely that Webster knowingly purchased or supported the production of fake artefacts. Despite the controversy surrounding his dealings with Leeds City Museum, and later disagreements with Robert Falla, Webster's dealings with the New Zealand Government and the museum were described as entirely honourable by other museum directors and curators.

Hei matau: pendants

The wearing of hei matau, or stylised Māori fish-hooks manufactured from bone, ivory or pounamu (nephrite) and tangiwai (bowenite) in the late 20th and early 21st centuries has become a symbol of Māori cultural revival and New Zealand identity.[287] This custom is partially based on European interpretation of artefacts worn as pendants, which included functional fish-hooks. Examples of pendants made from greenstone or bone were identified as hei matau or stylised fish-hook decorative neck pendants in the early 1900s.[101, 288] Some of these hei matau, referred to as porotaka hei matau[35] because of their rounded shape (Fig.60), to distinguish them from true hei matau (Fig.61; Fig.83) may not be stylised fish-hooks or stylised items, but represent traditional tools, and knowledge of their original function has been lost to present generations.[287]

In 1893, Director of the Colonial Museum Augustus Hamilton noted that Māori, as reported by Captain Cook and other early voyagers and missionaries, were not in the habit of wearing elaborate necklaces. He observed that although numerous examples of greenstone pendants are represented in museum collections, necklaces made from strings of perforated stones were virtually unknown and did not come into fashion with the adoption of European customs

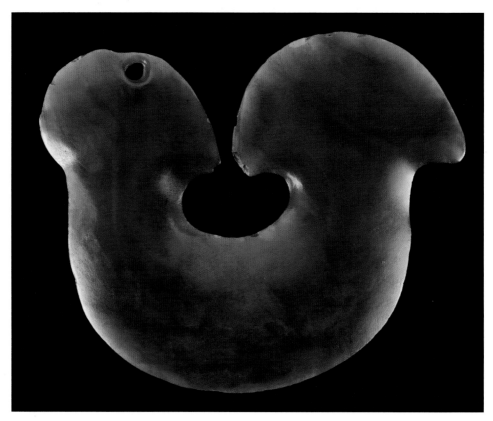

FIG.60 **Porotaka hei matau. 83 x 105 mm. Date unknown. Museum of New Zealand Te Papa Tongarewa, Wellington, OL000096**

and habits,[288] although strings of human, shark and dolphin teeth, stone and/or ivory 'reels', sections of albatross bone and tusk shells (*Antalis nana*) were worn.[20, 21, 36, 249, 289-291] The majority of ornaments documented as being worn by Māori in the 1770s include albatross down, perforated human teeth, sharks' teeth, greenstone kuru (ear pendants) and tiki. The popularity of kuru rapidly increased in the late 1700s, possibly as a result of demand from European trade interests.[292, 293]

Early explorers and traders in taonga (artefacts) exchanged European metal tools and weapons for traditional Māori wood, bone, shell and stone (particularly greenstone), implements and pendants,[249, 263] many of which are now in museum collections around the world.[294] Henry Skinner made a comprehensive study of Māori amulets[249, 289-291] in the early 1900s and reported that most were singular and had a wide variety of forms, including representations of bats (pekapeka), lizards (moko), fish (ika), human (tiki) (Fig.62), coiled eels (koropepe) and one-piece fish-hooks (matau).

Elsdon Best noted that neck ornaments were limited in range and reported hei tiki, koropepe and pekapeka,[35] commenting that, "inasmuch as Māori garments lacked pockets, it was quite a common practice to carry small implements or tools suspended from the ears or on a string around the neck". Best noted that small stone chisels were often carried this way, and if the chisel was made from greenstone, then the pendant was viewed as a desirable adornment apart from its usefulness. This tradition of wearing valuable tools, including fish-hooks, for safekeeping and decoration, was widespread throughout Polynesia and the wider Pacific. A wide range of tools that were worn as neck pendants has been illustrated, including poria (bird leg tethering rings) and fish-hooks (hei matau), and a flat, wide, circular pendant that was described as porotaka (rounded) hei matau.[26, 53, 185, 295] Some were fashioned from bone, ivory, or common stone, though most were of greenstone.

Hei tiki and other greenstone pendants may have been rare and crudely made until the availability of diamond-tipped drills after European contact made manufacture easier.[292, 293, 296, 297] Skinner noted how realistic looking, but fake pendants, handled by a generation or two of Māori could impart a characteristic silkiness to the surface and be transformed into "genuine" artefacts, and that many faked amulets in nephrite or bone were made by experts in Auckland, the most skilled being Trevor Lloyd.[298]

Historical records clearly show that hei tiki and kuru forms were pre-European, but there is no historical evidence for some of the rarer and more distinctive ornaments, including hei matau.[36, 290-293] Archaeological evidence for the use of matau as a necklace ornament was first reported in 1893 from a South Island site near Lake Manapouri.[288]

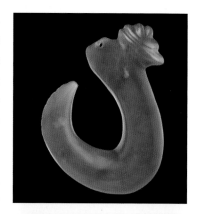

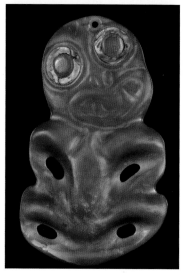

TOP FIG.61 **Pounamu (greenstone) fish-hook with a groove for attaching the fishing line, as well as a secondary hole to suspend the hook as a pendant when not in use. 60 x 46 mm. Date unknown. Ngāi Tahu, Kaiapoi. Museum of New Zealand Te Papa Tongarewa, Wellington, ME000608** LOWER FIG.62 **Hei tiki pendant in human form 114 x 72 mm. Date unknown. Museum of New Zealand Te Papa Tongarewa, Wellington, ME002972**

Following European contact, the perceived superiority of metal for working implements quickly became apparent and redundant stone, wooden or bone tools as material symbols of Māori culture were soon discarded, particularly in the period from 1790 to 1840, when

Authentic, replica or fake?

Māori fish-hooks were obtained from archaeological sites, ethnographic acquisitions by early explorers, or they have contemporary origins – traded in commercial exchanges. The non-archaeological hooks in museum or private collections may be genuine or misrepresentations: imitations, replicas, reproductions, fakes or even forgeries.

Commercial replicas of traditional artefacts are generally regarded as fakes or inauthentic. The demarcation between 'inauthentic' and 'authentic' fish-hooks however, is dependent upon the intent of the maker or seller to deceive, not the hook itself. Replica hooks made for trade that were passed off as genuine hooks made for fishing are the most common fraud. These cultural substitutes helped broaden access to rare taonga and relieved pressure on the worldwide demand for the acquisition of antiquities and ethnic art.

In the 19th and early 20th centuries Māori replicated or copied traditional designs in the production of fish-hooks for sale; therefore these hooks can be regarded as 'authentic'. Many of these trade items were sold to dealers who were often well aware that the hooks, although authentic, had been produced on demand and not for fishing. Dealers then onsold the hooks to unsuspecting collectors and tourists, fraudulently describing them as original examples of traditional fishing artefacts.

European forgers and lapidarists made fish-hooks for sale and often employed Māori to hand-polish the inauthentic semi-finished hooks; the resulting 'patina' was similar to that of genuine artefacts. Māori also sold these fake replicas (frequently on commission from dealers), to unsuspecting tourists. Today, although some fake hooks can be clearly identified, without

numerous sealing and whaling expeditions visited the coast.[37] Through the introduction of steel tools, carving was rendered easier but it also became elaborate and ornate, which spoiled artefacts intended for actual use, but improved them for trade purposes.[37]

The rapid adoption of metals by Māori meant that traditional items were either traded for metal implements or simply thrown away.[37] In the generally acidic New Zealand soils, flax lashings and wooden components quickly decayed,[1] while bone and shell components may have persisted for a few years or even decades, but ultimately they too decayed, leaving only stone components such as adze heads and minnow shank lures.[37, 223] Valuable items made of greenstone were often kept and worn as personal items or traded, and the original function of the tool was not always noted by subsequent generations.

Museum collections contain many examples of cultural artefacts catalogued as hei matau.

details of the maker's intent it is impossible to determine if many fish-hooks are genuine, authentic artefacts, or inauthentic replicas, copies or fakes.

Misappropriation of original designs or cultural artefacts by forgers, the acceptance of imitations or replicas and substitutes, and opportunists taking advantage of traditional culture is intellectual property theft. Interpretations by indigenous artists are the exception: their stylised designs are firmly based on historical prototypes and are intended to pay tribute to ancestral cultures. The fish-hook itself is reconstructed, and represents the ongoing development of a living and dynamic culture.

FIG.63 **Replicas made for trade often lack the inturned point of functioning traditional hooks but without knowledge of the maker and their intent it is impossible to classify the hook as authentic or fake. Date unknown. Museum of New Zealand Te Papa Tongarewa, Wellington, ME012119**

These forms have been interpreted as stylised fish-hook breast pendants[36, 53, 101, 140, 290, 291] which are distinctive but are unrelated to other breastplate discs that resemble pearl shell discs found elsewhere in the Pacific.[299]

In the late 1800s, the demand for curios from European collectors increased and the more ornately carved items were particularly sought after. By this time, most original fish-hooks had been sold or discarded. In order to meet this new demand, many hooks were manufactured by Māori. These replica hooks, although authentic, were made solely for trade purposes.[2, 26, 140, 263, 290, 291] European lapidaries produced fakes or inauthentic hooks: imitation or decorative greenstone hei matau using the spiral fish-hook pattern in large numbers, and other fakes made from bone, were also common.[290, 293]

Sir Peter Buck observed that hei matau pendants made using the incurved spiral design (koru) were rare, but appealed strongly to European makers of greenstone 'curios' and noted that frequently European manufacturers of fakes would employ Māori on commission to sell their "valuable family heirlooms" to unsuspecting collectors.[37, 300] Māori were shrewd entrepreneurs and it is clear that many hooks were not intended to catch fish; rather they were to catch the eye of the Europeans. The hooks, which became increasingly ornate, are examples of a formerly rare category of taonga which came to be clearly specifically designed and produced for their desirability as trade items, mirroring a similar process of the most internationally identifiable Māori motif, the hei tiki.[297]

Porotaka hei matau

In the early 20th century historians described a peculiar wide, circular hei matau pendant that was alluded to as a *porotaka hei matau* on account of its form. Many of these circular taonga, catalogued as hei matau in museum collections, lack a number of features characteristic of matau, traditional Māori fishing hooks.[2, 287] Ethnographic reports of Māori pendants in the early 1900s noted that greenstone tools were often worn as personal ornaments,[35, 53, 290, 291] but none recorded use of porotaka hei matau as tools.

Matau, particularly those manufactured entirely from stone such as pounamu, were made to a circle-hook design.[2] Although described as stylised fish-hooks, porotaka hei matau appear to have had a completely different function.

Traditional fishing matau can be distinguished from porotaka hei matau, as they are rounded in cross-section to provide strength and were manufactured with a groove at right angles to the direction of the point as an attachment for the fishing line. Some valuable greenstone matau have a hole drilled at the top of the shank in addition to the groove, and were suspended on a string and worn as personal adornments when not being used, as a means of safekeeping.[35] The suspension holes of porotaka hei matau are generally well worn and many examples have a second hole, apparently drilled after the first was worn through, indicating that they are of considerable age, but they lack the angled groove characteristic of functional matau.

Porotaka hei matau are predominantly found in southern areas, but are known from all regions of New Zealand. Provenance details of many early examples in museum collections do not have locality data, and do not distinguish examples found in archaeological sites from those traded by Māori. Some may even have been made in European lapidaries based in Dunedin and Auckland, or even in Berlin.

The similarity of porotaka hei matau throughout New Zealand suggests a functional purpose, resulting in consistency of the design (unless they were made only in certain areas before being traded). If porotaka hei matau were 'stylised', it would be expected that the design would vary from district to district (and over time), without some contemporary method of passing on

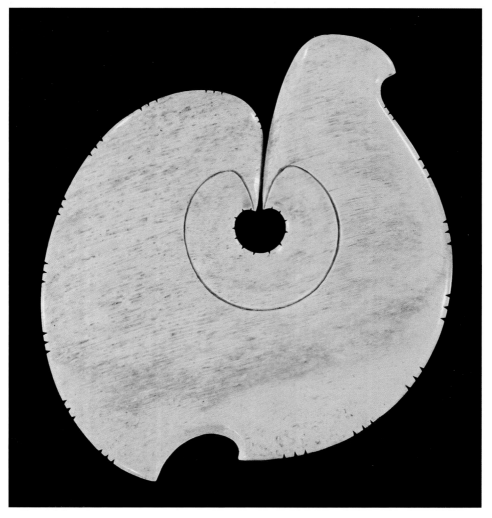

FIG.64 **Contemporary whale-bone hei matau. Rangi Kipa 1998. Museum of New Zealand Te Papa Tongarewa, Wellington, ME016927**

design being available. Significantly, no karakia associated with hei matau have been recorded.

Rather than being variable and stylised, traditional porotaka hei matau show a consistency in design and are U-shaped, with a broad outer leading edge and broadly notched shanks at either end, one or other of which has a drilled suspension hole. The internal section generally has a narrow gap, flanked by a shallow notch on one or both sides, which only superficially resembles the double internal barb (kaniwha) design of much smaller (~25 mm) fish-hooks that are rounded in cross-section. Porotaka hei matau are generally large (80-110 mm wide) and are flattened in cross-section with a sharp leading edge,[287] suggesting a functional rather than purely stylistic purpose.

Augustus Hamilton suggested that these porotaka hei matau were used in some manner for cutting hair along with a piece of sharp obsidian, but there is no direct testimony from Māori on this. Indeed, it is unlikely that anything valuable would be used in the haircutting as Māori culture demanded that it would have to be destroyed after the process.[101] The haircutting ceremony was extremely tapu but observers in the literature have made no mention of the use of haircutting tools other than sharp flakes of obsidian or sandstone.[15, 301] It seems unlikely that a haircutting artefact would be worn and openly displayed as an ornament.

If porotaka hei matau represent tools, and their function was replaced by the use of steel implements shortly after European contact, we can only speculate on their purpose. For example, the overall shape of the outer edge of porotaka hei matau is similar to the outer curve of the green mussel shell, used by Māori to prepare flax for net-making or weaving. The large (palm-size) pre-European porotaka hei matau may have been manufactured specifically for a task such as scraping flax, and would have been of particular value in inland regions or where mussel shells could not be easily obtained, such as areas without rocky coastal reefs. The internal opening with notched ends could have functioned as a gauge, ensuring that strips of flax required for net-making or other purposes were uniform in width; therefore the similarity of form throughout New Zealand may have been achieved through sharing of common knowledge. Alternatively, their predominance in southern areas may be associated with scraping fat from seal skins, or they may have been used throughout New Zealand for scaling and preparing fish.

Today, contemporary porotaka hei matau are purely ornamental, have no known functionality (Fig.64), and show considerable stylistic variation. It can be speculated that the pre-European porotaka hei matau may have been made for the purpose of scraping and shredding flax, preparing fish, or even for scraping seal skins, but the knowledge of their true function was lost following the introduction of steel tools after 1769.

Such conclusions are based solely on ethnographic accounts and require further study and interpretation. These large greenstone porotaka hei matau, manufactured by pre-European Māori, lack characteristics found on matau and are not stylised fish-hooks. Hence the term porotaka hei matau for these taonga is misleading and they probably represent artefacts in their own right, which were kept and worn as adornments when no longer required as tools.[287] A comprehensive study of porotaka hei matau in museum collections (including microscopic examination of wear-marks) is required, and particularly of examples collected prior to the late 1800s.

Hei matau pendants denote the importance of fishing to Māori, and their relationship to Tangaroa (the guardian of the sea and its environs). Decorative hei matau are considered to be symbolic representations of the fish-hook used by the ancestral Polynesian cultural hero Māui, who, according to metaphorical narrative, hauled up the North Island of New Zealand, Te Ika a Māui, from the depths of the ocean during a fishing expedition with his brothers.

The interpretation by Europeans of porotaka hei matau as stylised fish-hooks is possibly incorrect, but this interpretation has influenced contemporary Māori culture, giving rise to the modern development of often highly stylised and elaborate hei matau. Although based on traditional Māori taonga, the integration of European ideas, and subsequent revival of the custom of wearing now-stylised porotaka hei matau, can be regarded as representing a more generic nationalistic culture within Aotearoa.

The vast majority of hooks collected by James Cook and other early explorers are plain and functional, without ornate carving or decoration,[229] but because of the loss of wooden components from the archaeological record the full extent of ornate carving of pre-contact fish-hooks is unknown. Early 'archaic' period Māori fish-hooks resemble hooks from other areas of Polynesia and can be distinguished from later 'classic' period hooks, which have more ornamentation and reflect a cultural change that began in northern areas of New Zealand.

The distinction and interpretation of both earlier and later carving styles relies largely on 18th and 19th century evidence and undated artefacts to construct a hypothetical sequence of change.[36, 302] The demand for artefacts by European tourists and collectors in the latter part of the 19th century resulted in the production of a large number of replica hooks that cannot easily be distinguished from many earlier pre-contact examples.[142]

The use of hei matau for personal adornment, and the loss of wooden and flax hook components from the archaeological record, complicate interpretation of the traditional Māori hook-and-line fishing technology, which became dominated in the late 19th and 20th centuries by cheap, mass-produced metal hooks.[287]

Whakatapeha: *artefact manufacture and trading*

"Few nations delight more in trading and bargaining than this people; a native fair or festival best illustrates this fact. To such an excess are the feelings of the people carried in bartering with each other, that during war, though the belligerent parties seek for the annihilation of each other, yet at intervals a system of trade, as we have already stated, is carried on, that can scarcely be credited by strangers to their customs... Any persons having dealings with them are aware of their passion for commercial pursuits."

Settler and merchant Joel Samuel Polack, 1838 (Polack 1838: 74)

Māori trading networks were well established throughout New Zealand prior to the arrival of Europeans. Valuable raw materials including obsidian (māta tūhua), greenstone (pounamu), and whale bone (hihi tohorā), were traded in exchange for fish or other resources between different regions of the country and between coastal and inland tribes. Māori use of natural materials declined rapidly after European contact as they eagerly sought new trading opportunities to obtain metal tools. Māori continued to make fish-hooks to the circle-hook design using metals, including wire, copper ships' nails and even horseshoes, well into the 1800s, until imported cheap, mass-produced metal hooks became readily available.

Māori demand for useful European products, particularly metal tools, effected changes in what Cook's crew offered over the course of the three voyages. On the first, they exchanged mainly 'trinkets' and 'trifles' such as ribbons, paper, beads, medals, nails and things acquired in Tahiti, including bark-cloth manufactured from the paper mulberry (aute, *Broussenetta papyrifera*), which resembled Māori textiles. These products were initially popular as novelties, but quickly fell out of favour through oversupply. While trading offshore near Tolaga Bay in 1769, Cook noted: "This kind of exchange they seem'd at first very fond of, and prefer'd the Cloth we had got at the Islands to English Cloth; but it fell in its value above 500 p. ct. before night..."[17, 303]

On Cook's second voyage, trade goods were brought: "…to be exchanged for Refreshments with the Natives or to be distributed to them in presents towards obtaining their friendship, & winning them over to our Interest…"[20] and were of a more utilitarian character. These included iron tools, including fish-hooks, as well as adzes, axes, hatchets, chisels, saws, augers, hammers, knives, scissors, tweezers, wire, nails, combs, looking-glasses, kettles and pots, grinding stones, old clothes and sheets.

For their part, Māori also prepared selections of items to appeal to European tastes. At Tōtaranui (the original Māori name for Queen Charlotte Sound) in 1777 the *Resolution* was approached by canoes offering curiosities specially assembled for trade, including necklaces made of human teeth, a woman's dancing-apron decorated with red feathers, white dog skin and pieces of pāua shell, as well as greenstone and ornaments. Even at this early stage, perceived demand may have initiated the production of customised artefacts by Māori for trade with Europeans.[303]

Māori traded with early explorers, sealers and whalers who were seeking provisions and services, and also provided supplies to newly established settlements. Māori began exporting goods by the early 1800s, including prepared flax, dried fish, potatoes and grain to the Australian goldfields.[304] In 1848 an editorial in the *New Zealander* newspaper, having affirmed that Māori were the largest purveyors of foodstuffs, added pointedly: "…so large indeed as nearly to monopolise the market and to exclude the European settlers from the field of competition…"[305]

In the 1830s Māori had begun to build and purchase trading ships to carry surplus produce to centres of European settlement and in 1859 the Rev R. Taylor commented on the astonishing number of small coastal vessels in Auckland's Waitemata Harbour laden with produce that belonged to Māori.[89, 305]

Replicas and forgeries

Following the influx of European settlers after the signing of the Treaty of Waitangi in 1840, European numerical superiority was achieved around 1860 and Britain passed over political control to the settlers. War over disputed land claims ensued and racial attitudes hardened. In the wake of the wars the New Zealand Parliament passed a series of laws – for example, the Oyster Fisheries Act 1866 and the Fish Protection Act 1877 – destined to break Māori control of the resources of the land and sea, and significantly, to put an end to their competitive trading habits.[92]

Excluded from commercial trade, Māori turned to other ventures, such as tourism, and the sale of souvenirs became a vehicle for economic recovery. Tourists had begun arriving in New Zealand from the late 1860s, and by the 1880s, areas such as the geothermal region around Rotorua and the southern fiords were popular destinations.[306, 307] The travel agents Thomas Cook promoted tourism from the late 1880s and their itinerary entitled *Wonderland: Wellington*

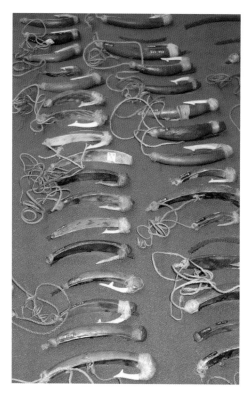

FIG.65 **Part of one of several drawers of museum pā kahawai held at Puke Ariki, New Plymouth.**

to Auckland overland included visits to Māori villages, where tourists could purchase artefacts and heirlooms directly from the tangata whenua.[300, 303, 308]

Demand for Māori artefacts by tourists and collectors[309, 310] in the late 19th and early 20th centuries soon outstripped the supply[140] and led to the manufacture of large numbers of replica fish-hooks.[141, 263]

Curio dealers sold Māori-made replicas to collectors in Europe and there is evidence that items ranging from hei kakī (pendants) to matau (fish-hooks) were being manufactured for sale at the Māori village of Parihaka in Taranaki around 1888 and were sold to a local New Plymouth dealer, James Butterworth of The Old Curiosity Shop.[263] One sales catalogue, produced in 1895 by Butterworth, included 400 pā kahawai lures and 176 other matau or suspended hooks.[263, 311, 312] As a result, many pā kahawai and matau now in museum collections are replicas made specifically for trade with the curio-hunting tourist,[140, 263] rather than for fishing (Fig.65, 66, 67).

New Zealand dealers also imported pseudo-artefacts or forgeries, including tiki, from Europe where they were manufactured, to supply the demand.[26, 37, 263, 298, 300] This practice continues today, with plastic and jade tiki being imported from Asia for tourist souvenir stores. Large numbers of Māori artefacts were made by known forgers, including James Frank Robieson, James Edward Little and others,[141] and many fish-hooks were supplied to European museums in the late 19th century.[141, 263] Many of those fish-hooks were obtained from Robieson between the years 1881-83 and are reported as being from Otago.

During this period the leaders of the Māori pacifist resistance movement at Parihaka, Te Whiti o Rongomai, Tohu Kakahi and some of their followers, were imprisoned without trial in Dunedin.[313] The government, together with Mr William Dickson, owner of a Dunedin lapidary and a known producer of replica greenstone tiki and mere, supplied the incarcerated Māori with pieces of greenstone and other raw materials for them to make items such as handclubs (mere), pendants (tiki) and ear pendants (kuru), which were exchanged for clothing and other

LEFT FIG.66 **Well made but insubstantial replica or fake hook. 260 x 180 mm. Museum of New Zealand Te Papa Tongarewa, Wellington, ME001409-81** RIGHT FIG.67 **Composite hook typical of the style of hook that was being made in the late 1800s. Date unknown. Collected by Harry Beasley between 1894 and 1939, British Museum Oc1944,02**

comforts.[16, 300] It is possible that contact between Robieson and Te Whiti at this time led to Robieson's commissioning Te Whiti to make fish-hooks as well as other artefacts for sale, and that subsequently Te Whiti continued the production of artefacts in collaboration with the dealer James Butterworth in New Plymouth[263] after Te Whiti's return to Parihaka in 1883 (Fig.67).

Many Māori fish-hooks made in the late 1800s have bone points that are slightly curved upward, and richly serrated or barbed, and are typically longer and more ornate than bone points found in archaeological sites. A bait string (pakaikai) was an essential component of the traditional hook. The thick wood or bone shank prevented bait being threaded onto the hook, as in modern metal hooks; hence the bait had to be tied to the lower bend of the hook, leaving the point and narrow gap free to trap the fish's jaw.

Many traditional hooks, particularly smaller one-piece bone hooks, had a small hole, notch or protrusion at the outer portion of the bend of the hook, to which the bait string was attached. Larger wooden composite hooks sometimes had a bait string extending from the snood lashing (whakamia). In most museum examples the bait string, when present, has been confused with the snood whipping string (whewheta) which was used to protect the lashing, and has been wound around the snood lashing as well. No known replica or fake hooks have bait strings attached.

Early collectors obtained numerous hooks which have been described as being made of human bone or with points of human bone, which seems to have enhanced the perceived value of the hook. Although human bone, particularly the bones of slain enemies, was used to make fish-hook points in pre-European times,[1] it is not possible to visually determine in many of the hooks if bone components are human, particularly if the bone has been polished.

At present, DNA sampling techniques are too destructive to allow for an adequate sample to be extracted and tested; hence, analyses must await less intrusive techniques. In the future, it may be possible to identify the materials used in the manufacture of traditional hooks through DNA analysis, and to distinguish modified or replica examples that use materials only available in the post-European contact period. Future development of techniques may provide some interesting insights, provided the issue of DNA contamination through years of handling can be resolved.

Dealers and trading

Manufacture of fish-hooks for trading purposes in the late 1800s may have been limited to areas where dealers were active. Wood-backed pā kahawai lures in museum collections are predominantly of 19th century form,[36] and appear to be restricted mainly to Taranaki, Northland, and perhaps Auckland and Thames, areas where artefact traders were active.[263] Māori artefacts are still highly sought after by present-day collectors, with auction house prices in 2011 for pre-contact pā kahawai and matau ranging from $700 to $3200 and $1000 to $5000 respectively. Post-contact examples range from $200 to $1200, although prices for some ornately carved hooks, and hooks with suggested 'Cook voyage' provenance, reach up to $12,000 (Fig.68).[281]

Many pā kahawai lures in museum collections are decorative rather than functional and typically have wooden shanks with bright, often highly curved portions of inlaid pāua shell, and the bone (occasionally greenstone, or even wood) points are often fragile and delicate, carved more as replicas of steel fish-hook barbs than the typical stout bone point of matau. Sir Peter Buck noted that the barbed points of these characteristic lures were an anomaly.[37] They are quite distinctive but have never been found in a reliable archaeological context.[44]

The inlaid pāua shell and delicate nature of pā kahawai lures were highly sought after toward the end of the 19th and early 20th centuries by traders including James Butterworth (Taranaki), Edward Spencer (Auckland), Sygvard Dannefaerd (Auckland and Rotorua), David Bowman (Christchurch) and Eric Craig (Auckland), as items for Victorian artefact collectors such as William Skinner (Taranaki), Andreas Reischek, Willi Fels, Augustus Hamilton, Alexander Turnbull, Thomas Hocken and Walter Buller, amongst many others.[141, 263, 314-317]

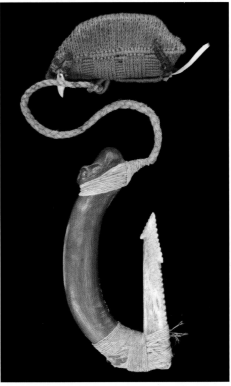

LEFT FIG.68 Māori fish-hook that 'may have' been collected during Cook's voyages to the Pacific, sold for $5500 at auction in New Zealand, March 2012. RIGHT FIG.69 Traditional Māori hooks made for fishing are characterised by their stout components without overly ornate bone points, and a quality of craftsmanship that is lacking in many later replicas. Composite hook with attached bait bag. Musée du quai Branly, Paris, 71.1978.50.1.1-2

Older pā kahawai were manufactured with shanks made from the thick concave rim of a pāua shell, with a simple non-barbed point (of shell, wood or shark fin spine), lashed to the distal end.[2, 140] Other straight-shank bone or stone lures, without pāua shell linings, had the point attached directly to the cord of the fishing line which extended down the length of the shank.

Examples of lures with pāua shell lashed to the inner side of a concave wooden or whale-bone shank are more recent. These latter examples were made with delicate sections of pāua shell inserted into wooden shanks – usually tōtara. The use of easily worked tōtara wood made it possible to manufacture numerous replica pā kahawai in a short period of time. Large numbers of these spinning lures were produced at the behest of the curio-hunting tourist and although many have been described as sad travesties of the older forms,[140] no reliable features to distinguish replicas or forgeries from authentic lures have been noted.

Many of these lures, known as 'museum pā kahawai'[36] show some degree of European influence in the frequent use of linen thread or sisal rather than prepared harakeke (muka) for bindings that are often crude and untidy (Fig.71) – traditional lashings were made with fine 2-ply twisted cord.[37] Some of these lures have copper barbs, most have signs of steel chiselling to create the delicate pāua inlay, and many are overly curved and would spin uncontrollably in water. Very few have any provenance details, and those that do, often include only details of collectors, rather than the origin of the hook itself.

Many replicas do not meet the design requirements of a rotating hook which allows them to function efficiently. In particular, the lack of inturned points or angled snood lashings; crude, often atypical adornment carvings, or overly ornate carvings; and the use of non-New Zealand plant fibres, are all indicative of hooks that were never intended to be used for fishing.

Māori artefacts gathered by James Cook and other members of the three voyages from 1769 to 1777 that are now in European museums are the oldest non-archaeological Māori cultural items in existence. The fish-hooks known to have been collected during the Cook expeditions and by other explorers in the late 1700s and early 1800s, are characterised by a high quality of workmanship in their construction and are stoutly made with careful and precise lashings of muka fibre (Fig.69, 70).

TOP FIG.70 **Rotating hook collected on the first Cook voyage. Georg-August-Universität Göttingen, Oz 327** LOWER FIG.71 **Museum pā kahawai have wooden shanks with delicately carved pāua shell inlays, but lashings are often crudely tied. Puke Ariki, New Plymouth, A57-783**

Hooks with carved ornamentation collected by early explorers (pre-1800), are extremely rare. An example of a pre-1800 composite hook with a detailed human figure carved on the shank is held in the National Museum of Ireland[11] (Fig.86), with another hook that has a small carved head on the snood knob;[12] both were collected on Cook's second or third voyages.[229] In his journal account of Cook's first voyage, Sydney Parkinson illustrated nine fish-hooks, but of these, eight are plain and only one large hook has an ornate carved head on the snood

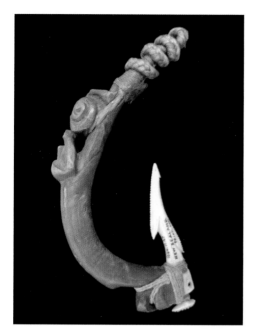

FIG.72 **Replica or forgery composite Māori fish-hook.** Pitt Rivers Museum, Oxford, 1884.11.47. Photograph by Chris Paulin.

knob (Parkinson: Plate XXVI, fig.6)[22] This hook, if it still exists, does not appear to be among any of the known Cook voyage fish-hooks in museum collections. All other Māori fish-hooks known or thought to have been collected during the Cook expeditions are plain; similarly, archaeological fish-hooks with carved adornments are rare. In contrast, many hooks obtained by collectors and museums in the late 1800s have ornately carved snood knobs, and some hooks obtained by museums in the early 1900s also have detailed carving on the shanks.[142]

No wood-backed pāua shell pā kahawai lures are represented in the collections made by early explorers, and the earliest known example dates from the late 1840s. It is possible that the detailed carving of bone and wooden fish-hooks, and the delicate pāua shell inlay required for pā kahawai, were not easily produced until steel tools became available. The production of many ornate hooks in the late 1800s and early 1900s was in response to demand created by European dealers and collectors.

Carving style has been used to divide Māori artefacts into different time periods.[302] The earliest Nga Kakano (the seeds), dates from AD 900-1200; Te Tipunga (the growth) from AD 1200-1500 marks the development of a distinct Māori style; Te Puawaitanga (the flowering) dates from AD 1500-1800; and Te Huringa (the turning) from when steel tools became available and carving became more elaborate and detailed, dates from 1800 to the present.

Māori fish-hooks probably from Cook's second and third voyages held in the National Museum of Ireland have been assigned to two different style periods, Te Puawaitanga and Te Huringa.[318] Other fish-hooks from the Oldman Collection (accumulated between 1890 and 1913), have been assigned to particular style periods,[302] but without provenance details many hooks cannot be identified as pre-European, or as 19th century replicas or forgeries, and the dating of hooks by carving style cannot be regarded as accurate.

---

[11] National Museum of Ireland AE1893-760
[12] National Museum of Ireland AE1893-761

FIG.73 **Composite Māori fish-hook.** © Trustees of the British Museum. British Museum, London, BM 9356

Although the vast majority of hooks and lures collected by early explorers in the late 1700s are functional and utilitarian, without ornamentation, it is possible that ornately carved hooks, and trolling lures with attractive pāua shell inlays, were collected by early explorers and were gifted to wealthy patrons, but remain in private collections rather than in public museums. Over 100 collections containing Māori artefacts are known to exist in Europe, but there is no complete catalogue of the taonga held.[13]

Carvings on hooks are mostly representations of heads rather than full figures. One composite hook with an almost complete figure carved on the shank is held in the Pitt Rivers Museum, Oxford, England[14] (Fig.72), and has been described as an important example of hooks that were formerly quite common.[140] Beasley noted that hooks with unnecessary ornamentation were nearly always illustrated in the accounts of early voyages, but he could only provide a single reference to Sydney Parkinson's illustration, and stated that he was unable to illustrate any number.

The Pitt Rivers Museum hook (which was acquired in 1884 and has no known provenance), has an unusual inverted carved figure on the shank with inserted shell eyes and a plaited snood of sennet rather than New Zealand flax, and has been illustrated in colour,[319] but mistakenly described as: "a decorated Māori cutting tool with an edge created from inset shark's teeth", a caption apparently derived from the information provided for a different PRM object.[15] The inverted carved figure is unusual and crude, raising the possibility that the hook is a copy or forgery: superficially it resembles another example held in the British Museum[16] (Fig.73), which also has no known provenance but was displayed in the museum gallery in the late 1800s.

---

[13] Arapata Hakiwai, Te Papa Tongarewa, *pers. comm.*, 2010
[14] Pitt Rivers Museum #1884.11.47
[15] Pitt Rivers Museum #1886.1.1161 (PRM catalogue notes)
[16] British Museum BM 9356

FIG.74 **Early New Zealand tourist guide books:** *Willis's Guide Book of New Route for Tourists: Auckland-Wellington, via the Hot Springs, Taupo, the Volcanoes & the Wanganui River* (Allen 1894) and *The New Zealand Tourist* (Bracken 1879).

Tourist souvenirs

Recreational travel emerged in New Zealand in the 1870s (Fig.74).[306-309, 320, 321] This was toward the end of the New Zealand Land Wars and had a stimulating effect on Māori material culture,[303] producing innovative forms that are now widely regarded as part of authentic indigenous culture. The influence of 'invented' traditions may be pervasive in anthropology and related disciplines, and may have been applied to many minority cultures throughout the world.[67, 322-324]

Māori artefacts, such as fish-hooks, carvings, woven bags and cloaks manufactured specifically for the tourist industry combined concepts from Māori identity, travel marketing and European ethnology, and have undergone traditionalisation in a manner similar to that suggested for Māori meeting houses. Meeting houses are now accepted as valued survivals of an earlier time and are even described as classic examples of Māori culture,[6, 325-327] even though archaeological evidence reveals that large carved meeting houses as described by explorers from the 1830s onward did not exist before the arrival of Europeans.[36, 328, 329]

Māori were quick to adopt new materials as well as metal tools, and they continued to make fish-hooks using metals, imported fibres and traditional materials. These post-contact hooks were made for both fishing (new materials), and for trade (traditional materials). Whether the tool used was made of stone, shell, iron or steel, has less to do with authenticity than with the speed of manufacture.[267]

Fish-hooks were not the only items made for trade with Europeans. Greenstone ear pendants (kuru) were uncommon in 1769-70,

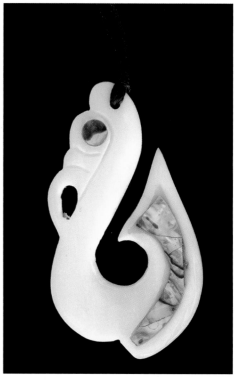

FIG.75 **Contemporary bone matau made in China and available as a New Zealand tourist souvenir.**

but increased enormously in response to European demand. Researchers have suggested that most of the greenstone Māori pendants in the Australian Museum are probably fakes.[292, 293] These objects are the products of dynamic indigenous situations which demonstrate initiative with new tools, ideas and materials, and they have been made with greater or lesser amounts of skill. The manufacture of fish-hooks as tourist souvenirs was not limited to New Zealand – it also occurred in other parts of the Pacific well into the early 20th century.[188]

The authenticity of Māori souvenirs representing cultural merchandise is significant – present-day souvenir artefacts that are not culturally meaningful are stereotyped as tourist kitsch (Fig.75). The presentation of Māori culture and souvenir artefacts has changed over time. Fish-hooks made after European contact followed traditional designs and used methods that are sufficiently conservative to provide clear, demonstrable links with the past, providing continuation and links to indigenous culture. This change did not significantly affect the practice of Māori culture, but can be seen either as promoting Māori culture and self-determination, or alternatively, as disempowering.[330]

Separating authentic from non-authentic artefacts is often couched in terms of whether something was made pre-contact and with stone tools (authentic), or made post-contact for sale and not for indigenous or traditional purposes (non-authentic).[267] Ethnological souvenirs from the late 19th and 20th centuries have an ambivalent status in museums.[303] They are valued for their age and exoticism, but where it is known that they were produced for the tourist market, they may be regarded as inferior examples of workmanship, and less valuable than those manufactured for native use. Through the influence of debates about the invention of tradition, souvenirs are often seen as hybrid and therefore inauthentic specimens of purely indigenous material cultural tradition, and are rarely knowingly exhibited in ethnological displays.[303]

Many hooks in museum collections are crudely made, and often incorporate non-traditional materials including linen thread and foreign fibres other than those from native New Zealand plants such as flax (harakeke), or cabbage tree (tī) for lashings. Other examples of non-traditional materials include using engineers' chalk to replicate the bone point, and hooks where the bone points are lashed to the wooden shank with wire, concealed by crude lashings of flax.[142] These poorly made hooks were collected in the late 19th and early 20th centuries, and few have good provenance details.

Today, decorative fish-hooks, intended to be worn as hei matau (pendants) are widely manufactured for sale to tourists, and are made locally using pounamu and imported jade by Māori and Pākehā carvers, as well as being imported cheaply from Asia.

8

Papa Tongarewa: *the European collections*

"One of the most important duties... of a new country is the formation of a scientific museum, the principle object of which is to facilitate the classification and comparison of the specimens collected in different localities... In this respect a scientific museum differs from one intended only for the popular diffusion of natural science – the former being a record office from which typical or more popular museums can be supplied with accurate information instructively arranged – an arrangement which would prevent their lapsing, as is too frequently the case, into unmeaning collections of curiosities..."

Colonial Museum Director James Hector, 1870 (Hector 1870: 1)

Cabinets of curiosities

Hooks collected by early explorers, including those from the Cook voyages, are of critical importance as they represent the first European contact from which identified collections were made, and thus provide a baseline of material from which studies can be undertaken and changes in material culture documented.[142]

The Pacific voyages of James Cook revealed to the Western world an entirely new vista of geographic and scientific knowledge. The first voyage was in fact a scientific mission, organised by the Royal Society of London to observe the Transit of Venus from Tahiti. The British Admiralty went to considerable lengths to ensure that each of Cook's voyages included learned men of science and their assistants and artists. Joseph Banks, Daniel Solander, Georg and Johann Forster, Anders Sparrman, Sydney Parkinson, Alexander Buchan, David Nelson, William Hodges, John Webber and others, whose primary interests were botanical and zoological (rather than ethnographical) all travelled with Cook.

The importance of the 'natural curiosities' collected on these voyages of discovery was recognised. The biological specimens gathered and described became type specimens for numerous

FIG.76 Ole Worm's cabinet of curiosities, from Museum Wormianum, 1655. Smithsonian Institution Libraries, Wikimedia Commons.

species and were well documented with notes on where and when they were collected.[294] After the ships returned to Europe, specimens that had been carefully described were deposited in museum collections or sold to willing buyers.

In contrast, although the scientists and other officers and crew made extensive collections and observations of ethnographic materials, the objects themselves were obtained as mementos and were often poorly documented and not highly regarded.[294]

In an address to the Dublin University Zoological and Botanical Association in 1856 the Chairman, Dr. Robert Ball, opened the meeting by saying:

"Tonight Professor Harvey favours us with some remarks on the inhabitants of the Fiji Islands, whose arms, etc., you see hung around the room. Collections of this kind have been sneered at but very improperly as a right knowledge of them is of great importance in the very difficult and very high study of ethnology; a study in which the utmost penetration of the zoologist should join with the most profound knowledge of the philologist as the races of men are not less distinguished by their physical form and language than by their arms and ornaments: these things have come to have a scientific use…"[331]

The Cook expedition members used the Māori curios they had collected to their own ends, willingly presenting them to royalty, admiralty, gentrified friends and learned colleagues. Others were gifted to patrons or private collectors in Britain, Germany and other European countries, or were simply sold at whatever profit they could get to collectors of artificial curiosities (Fig.76). As the various collections were sold or dispersed, the artefacts found their way into private cabinets of curiosities across Europe, and eventually, from there to public museums.[142] More than 2000 ethnographic artefacts, including dozens of fish-hooks, were collected during Cook's three Pacific voyages.[294, 332] The popularity of Cook artefacts in the 19th century, and subsequent extensive trading between collections, led to items being dispersed widely throughout Europe and after 200 years of curio trading relatively few fish-hooks can be traced directly to Cook, or even to the Cook voyages.

Many fish-hooks in collections that were donated or sold to museums in the late 19th and early 20th centuries (e.g. Beasley, Buller, Turnbull and Oldman collections), were made for the curio-hunting tourist or collector, and are replicas or even fakes (Fig.77). Because of this trade, there are few Māori fish-hooks in New Zealand museum collections, other than archaeological examples, which can be reliably interpreted as traditional hooks made for fishing. Few archaeological examples have retained any wood or flax components[36] and are mostly single-piece bone or shell hooks, or stone shanks, and unassociated bone points and fragments.

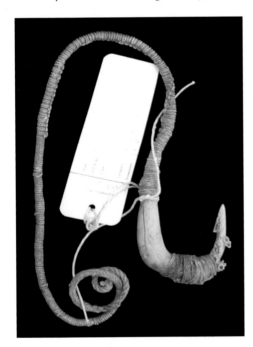

FIG.77 **Fake composite fish-hook made with a point from a trolling lure. British Museum, London, BM 1914**

Some Māori hooks in museum collections have been modified historically by collectors and museum staff for display and research purposes. As collections were established in the late 18th and early 19th centuries, there was a desire to display ethnographical items, including artefacts representing fishing technologies from distant lands, as 'artificial

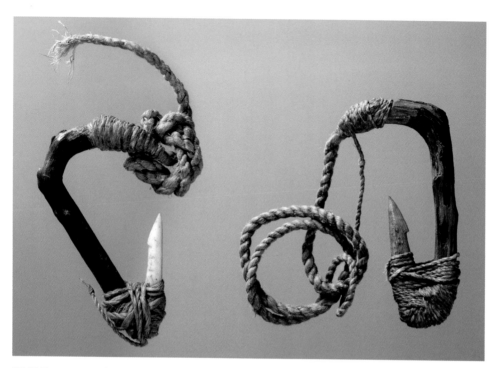

FIG.78 **Two composite hooks from the Forster Collection (Second Cook voyage). The point lashing of one appears to have been unravelled and re-tied. Georg-August-Universität Göttingen, Left: Oz 332; right: Oz 333**

curiosities'. Rather than preserving them as cultural artefacts in their own right, the collectors regarded the curiosities as part of the diversity and complexity of nature.[333]

Consequently, in order to display many fish-hooks, lines made of non-traditional materials such as hemp, sisal, jute, linen flax, cotton and even sennet were frequently added, or used to replace deteriorating lines. The added lashings were often incorrect and misleading. At least one hook known to have been collected during Cook's second voyage appears to have been modified by unknown collectors or museum curators; two composite hooks in the Forster Collection at Göttingen[17] are similar, and were possibly made by the same person. The condition of the hooks and catalogue description of how the bone point of the hook was inserted into a groove at the end of the shank suggests that the lashing of one has been unwound for examination, and then re-tied – the lashing is crude and unfinished when compared with the second (Fig.78).

Snood lashings that have been added to Māori hooks by collectors or museum curators, and lashings on hooks that have been made as replicas by those unfamiliar with functional Māori fish-hooks, are frequently tied so that the snood is aligned with the shank, parallel to the

---

[17] Georg-August-Universität Göttingen Oz 332 and Oz 333

FIG.79 **Albatross hook.**
112 x 66 mm. Museum
of New Zealand Te Papa
Tongarewa, Wellington,
ME005033

direction of the point. The rotating manner in which the traditional fish-hook functioned requires the snood lashing to be at an angle to the direction of the point (ideally at 90°), unlike metal J-shaped hooks in which the line is attached parallel to the direction of the point. Composite hooks intended for catching albatrosses and other seabirds are generally much lighter in weight than hooks of similar size made for catching fish (Fig.79). These composite hooks have snood lashings that are more in line with the point; however, it is not always possible to determine if some hooks with parallel snood lashings were made as albatross hooks, or were intended for sale as replica fish-hooks.

There are significant collections of early Māori artefacts, including many fish-hooks, in museums in Britain, France, Austria, Germany, Italy and Russia.[294, 332] In addition, over 100 other collections of Māori taonga (treasured

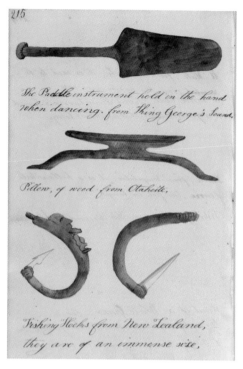

FIG.80 **Fishing hooks from New Zealand illustrated in Kenelm Digby's** *Naturalists Companion.* **New South Wales State library, Sydney, a155030**

artefacts) are known to be held in Europe, but the details of items held are largely unknown.[267] The most popular items obtained by early collectors were primarily smaller taonga made from wood, bone or stone, as well as carved and woven taonga. Once obtained from Māori sources, these items were easily transformed into Western objects of academic and monetary value, not only because of the exotic material from which they were made, but because of their easy portability.[334]

Many traditional fish-hooks in museums are poorly documented. This is a result of collections being accumulated as artificial curiosities or souvenirs rather than systematic attempts to preserve traditional artefacts or document material evidence of Māori culture. The date of collection of many hooks can be broadly established through cataloguing dates and known details of donors, but many hooks passed from collector to collector and most original details have been lost. The loss of provenance data and associated cultural significance[335] resulted in many collections becoming meaningless assemblages of curiosities.[336]

The often haphazard composition of late 18th century and early 19th century European museums reflected the then widely held belief that the diversity and complexity of nature was positive proof of the existence of a Divine Creator. This encyclopaedic approach is well demonstrated by Kenelm Henry Digby's *Naturalists Companion* prepared from specimens in the museums of Trinity College and Dublin Society in Ireland in the early 1800s,[333] which includes illustrations of two Māori fish-hooks from Cook's second or third voyages (Fig.80), besides numerous illustrations of a wide variety of animals and birds.

Digby's stated intention was to highlight to all but the most insensible mind, wonder at the formation and the various properties, and dispositions of the Brute Creation. Comparison of Digby's 1810-17 manuscript with published catalogues from early museums, such as the Leverian Museum or William Bullock's Museum, shows how close Digby's work was in conception to the layout and interpretation of contemporary museums.[333, 337]

New Zealand museums

Public museums in the provinces of Auckland, Wellington, Canterbury and Otago were established in the latter part of the 19th century. Although these museums hold extensive collections of archaeological fish-hooks, the comparatively late dates of their establishment resulted in few examples of ethnographic fish-hooks that can be reliably dated as pre-European. Many hooks that were donated to the collections have good provenance details of previous owners and collectors, but few have details of their original sources, and most were obtained after 1870.

Initially the museums developed from small private natural history collections and included examples of foreign fauna and flora, as well as foreign artefacts that were obtained through

FIG.81 **Colonial Museum, Wellington, watercolour by George O'Brien, 1865. Museum of New Zealand Te Papa Tongarewa, Wellington, 1992-0035-2275**

overseas exchange of New Zealand 'curiosities' such as moa bones, bird skins and Māori artefacts.

The Colonial Museum in Wellington (Fig.81), was founded with collections started in 1851-59 by the Wellington Philosophical Society and transferred in 1865. The Canterbury Museum, Christchurch, was first opened in 1870 and the collections were built up by its first director, Julius von Haast, Surveyor-General of Canterbury from 1861 to 1871.[338] The Otago Museum, Dunedin, was established in 1868 following display of provincial geologist James Hector's collection of 5000 rocks and minerals at the trade-promoting New Zealand Exhibition in Dunedin in 1865,[339] and the Auckland Museum was established in a farm worker's cottage in 1852 with a display of wool. Interest dwindled and in 1869 the somewhat neglected and forlorn museum was transferred to the care of the Auckland Institute, but remained relatively inactive until the early 1920s.[340]

The New Zealand Institute, now the Royal Society, was established in 1867 to co-ordinate and assist the activities of a number of regional research societies including the Auckland Institute, the Wellington Philosophical Society and the Otago Institute.[341]

FIG.82 **Pohau mangā held in the Museum of New Zealand collections has a tentative provenance from one of Cook's voyages. 135 x 45 mm. Museum of New Zealand Te Papa Tongarewa, Wellington, ME002494**

At least eight Māori fish-hooks in the collections of the Museum of New Zealand Te Papa Tongarewa have tentative links to the Cook voyages: two pohau mangā or barracouta lures[18] (Fig.82), two wooden composite hooks[19] (Fig.8), three internal-barb hooks[20] (Fig.30) and one top shell (*Coelotrochus* sp.) hook,[21] but this cannot be confirmed. The hooks were possibly among Cook voyage artefacts acquired by the collector William Bullock from the sale of the Leverian Museum collection in London in 1806, or were among other items that were given to him by Sir Joseph Banks.

In 1819, Bullock sold his entire collection; many items (including an Hawai'ian feather cloak and helmet) now in the Museum of New Zealand Te Papa Tongarewa, were bought by

[18] Museum of New Zealand Te Papa Tongarewa ME002494; ME002495
[19] Museum of New Zealand Te Papa Tongarewa ME002496; ME002497
[20] Museum of New Zealand Te Papa Tongarewa ME002498; ME002499
[21] Museum of New Zealand Te Papa Tongarewa ME002500 (not located)

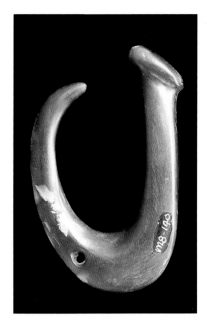

LEFT FIG.83 Valuable greenstone hooks were highly prized and rarely traded – none are known from the Cook voyage collections. Rotorua Museum, M8-140 RIGHT FIG.84 Fishing line of prepared flax (muka) with bone fish-hook and quartz sinker. Possibly of Cook voyage origin. Museum of New Zealand Te Papa Tongarewa, Wellington, ME012118

Charles Winn (1795-1874) for his private collection.[22] In 1912, after they had been in the family's possession for nearly 100 years, they were donated to the 'Dominion of New Zealand' by Charles Winn's grandson, the second Lord St Oswald.

There is no documentation associated with any of the hooks that can demonstrate direct links with the Cook voyage material. The Bullock sale catalogue notes indicated that many items were foreign curiosities 'principally' brought from the South Seas by Captain Cook. Bullock had also received items from other expeditions.[23] Unlike the larger and spectacular items, such as the Hawai'ian cloak, which are well documented, the fish-hooks cannot be conclusively associated with the Cook voyages, although they were collected prior to 1805.

A fourth one-piece internal-barb hook with an attached quartz sinker in the Museum of New Zealand Te Papa Tongarewa collections[24] (Fig.84) is almost identical to a hook and sinker from one of the Cook voyages held in the Museum of Anthropology and Archaeology, Cambridge, England (Fig.85). The Cambridge hook is richly decorated with red feathers from kaka (bush parrot, *Nestor meridionalis*), and has previously been incorrectly labelled as 'Sandwich Islands'

[22] Museum of New Zealand Te Papa Tongarewa archives MU000016/001/0016
[23] Museum of New Zealand Te Papa Tongarewa archives MU000016/001/0016

but the style of the hook and presence of kaka feathers confirms it is from New Zealand.[142, 294] The Museum of New Zealand Te Papa Tongarewa example lacks decorative feathers and was a gift from the Imperial Institute, London, 1955, when the Institute gave the Museum a significant collection of items associated with Cook. These had been in the possession of Queen Victoria (1819-1901) and had been given to the Institute by Edward VII (1841-1910). Cook himself may have given these to George III (1738-1820) after his second voyage; however, there is no documentation linking the hook and sinker directly to the Cook items.[25]

European museums

Only two museum collections in Europe, the Museum of Archaeology and Anthropology in Cambridge, England and Georg-August-Universität in Göttingen, Germany, contain Māori fish-hooks that can be positively attributed to the Cook voyages of 1768-79. There are many other hooks in European museums that were possibly collected during Cook's voyages, but their exact status cannot be verified as precise details of the collectors and dates have been lost.

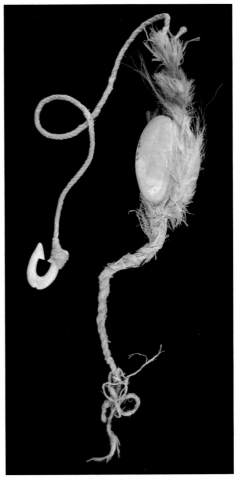

FIG.85 **Internal barb hook with quartz sinker decorated with kaka feathers. Probably of Cook voyage origin. Museum of Anthropology and Archaeology, Cambridge, 1925.365**

Collections at the Museum of Archaeology and Anthropology, Cambridge University, England, are especially important because much of the material was collected by James Cook himself, or was presented by Cook to John Montagu, the fourth Earl of Sandwich (1718-92), First Lord of the Admiralty, and a great supporter of Cook's. Montagu in turn presented the objects to Trinity College, Cambridge, where he had been a student (1735-37), and the College

---

[24] Museum of New Zealand Te Papa Tongarewa ME012118
[25] Museum of New Zealand Te Papa Tongarewa archives MU000016/001/0016

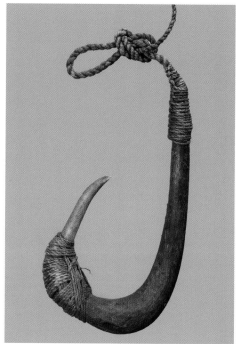

LEFT FIG.86 **Composite hook from the second Cook voyage. Humphrey Collection no. 245. Georg-August-Universität Göttingen, Oz 338** RIGHT FIG.87 **Hanover Museum composite hook. Date unknown. Georg-August-Universität Göttingen, No. 35**

deposited the collection in the museum in the early 20th century. The museum has 215 objects that can be traced to Cook's voyages, and about 50 Māori fish-hooks including several from the Cook voyages.[294, 342]

The collection also includes six composite hooks that are wooden with bone points, including one slender hook of the style used for catching seabirds.[26] One barracouta lure (pohau mangā) was collected by Cook and given to David Pennant by Joseph Banks, who in turn donated it to the museum.[294, 343]

Twenty-one pāua shell pā kahawai trolling hooks are represented in the collection, with one whale-bone shank example, three with metal backing, and the remainder wood-backed. Six have kiwi feathers attached at the distal end. One bone shank and bone point trolling lure is clearly fake.[344] The earliest example is dated 1853, and none were collected by Cook. Several other one-piece bone and shell hooks, and a selection of archaeological bone points were obtained on exchange from New Zealand in the early 20th century.

The Cook-Forster collection at Göttingen, Germany, represents one of the world's most

---

[26] Museum of Anthropology and Archaeology #1935.1266

distinguished collections of ethnographical artefacts from the South Pacific. Items were bought on commission by George Humphrey (a London dealer) for the King of England (George III), to be sold to Göttingen in 1782, and in 1799 the University bought the remainder of the personal collection of the deceased Reinhold Forster who had accompanied Cook on his second voyage (1772-75). The fish-hooks in this collection derive from Cook's second and third voyages (Fig.86), and some may have been purchased at an auction of the collection of David Samwell, who served as surgeon's mate on the *Resolution* from February 1776 to August 1778, when he was transferred to the *Discovery* to replace William Anderson, a surgeon who had died on the voyage in the Bering Sea.[345]

Samwell's collection was sold in June 1781, and in the only known annotated copy of the sale catalogue, George Humphrey is shown to have bought a number of lots, which probably became part of the collection prepared for Göttingen.[345] It is also possible that some of Cook's voyage items were obtained from Jacob Forster,[345] a relative of Reinhold and Georg Forster, who was married to Elizabeth Humphrey, sister of George Humphrey.[346, 347]

One composite hook[27] (Fig.87) was presented together with other pieces to the newly founded Hanover Museum in 1854 as part of the Cook-Forster collection. In contrast to other fish-hooks from New Zealand which are well documented, this hook is not mentioned in the Reinhold Forster legacy and cannot be identified conclusively from George Humphrey's catalogue.[348] The hook has a snood lashing that is almost parallel to the point, rather than at a strong angle, and resembles replica hooks made in the late 1800s rather than authentic hooks collected during Cook's voyages. It is possible that this fish-hook was originally a gift from Georg Forster to Reinhold Friedrich Blumenbach, the curator of the Academic Museum of Göttingen (besides other gifts given by Forster during visits to Göttingen in 1778 and in following years) or it was given to Blumenbach by Joseph Banks after his return from Cook's first voyage (Blumenbach and Banks corresponded for a long time).[348] Hence, that hook was included in the Cook-Forster collection. The type of snood lashing and the point are unusual for an early Māori fish-hook and it is possibly either a hook intended to catch albatross or a mid-19th century replica, and thus may or may not be part of the Cook-Forster collection.[142]

Hooks known to have been collected on Cook's second and third voyages in the collections at the National Museum of Ireland, Dublin, have become amalgamated and confused with hooks collected later in the 19th century.[318, 331] Although two hooks illustrated by Kenelm Digby in 1810-17[333] are possibly Cook voyage examples,[349] no documentation exists to connect them to Trinity College and Cook's voyages.[28]

---

[27] Georg-August-Universität Göttingen #35

[28] *pers. comm.* Rachael Hand, Curatorial Assistant, Museum of Archaeology and Anthropology, Cambridge

On Cook's second voyage, James Patten of Ulster sailed as surgeon on the *Resolution* and later settled in Dublin. During the voyage he made a collection of Pacific artefacts, which included several Māori fish-hooks, and presented it to Trinity College, Dublin, around 1780. Another collection in Dublin came from Captain James King who sailed on the third voyage, and took over when Captain Clerke of the *Discovery* died, after the death of Cook in Hawai'i.[318, 331] The items collected by King were not presented to Trinity College until after his death in 1784, and were donated by King's father, the Rev James King, who was Dean of Raphoe in the County of Donegal.

The Dublin Marine Society also donated further 'curiosities' to Trinity College in 1792, which had been collected on one of Cook's voyages.[331] No complete catalogue of the objects from Trinity College exists, so it is not possible to distinguish hooks collected by Patten from those collected by King or donated by the Marine Society.

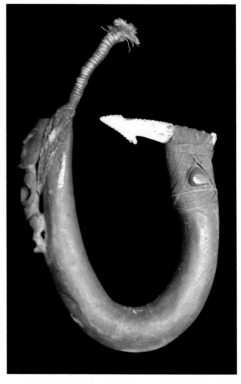

FIG.88 **A unique composite Māori fish-hook with a carved figure on the shank. Probably from one of Cook's voyages. National Museum of Ireland, Dublin, AE 1893 -760**

The items from Trinity College were transferred to the National Museum of Ireland in 1882 and 1885. In 1909 the museum purchased a collection of Māori artefacts from Dr Isaac Usher, who had acquired them from his father-in-law Captain George Meyler, who had fought in the Land Wars while in New Zealand between 1860 and 1889, and further items were added by travellers such as Dr James McKellar, and from other collections donated to the Royal Dublin Society and the Science and Art Museum, which has now become the National Museum of Ireland.[318]

The collection of Māori fish-hooks held in the National Museum of Ireland includes composite fish-hooks and lures – including wooden pohau mangā with bone points, and pāua shell pā kahawai. While some of these hooks and lures were obtained from the Cook voyages, others were collected in the late 1800s,[142] but the many collection items from New Zealand were not clearly labelled and became mixed during re-organisation of the collections in the early 20th century.

One composite wooden hook with a bone point[29] (Fig.88) is of particular interest. It is stoutly made and has a detailed carving of a full human figure on the shank. This hook appears to be one of two illustrated in the early 19th century by Kenelm Digby[333] (Fig.80). The second hook illustrated by Digby, also a composite wooden hook with a bone point,[30] has a carving on the head of the shank, although the illustration is poorly executed. The date of Digby's publication indicates that these two hooks were probably collected during the Cook voyages.[229]

On the third voyage, Cook's ships the *Discovery* and *Resolution* were re-supplied at Kamchatka, Russia, in the North Pacific during their unsuccessful search for the Northwest Passage in May 1779 (it was from Kamchatka that the news of Cook's death was conveyed to Europe). As a consequence of the assistance provided by the Russian authorities to members of the voyage, Captain Clerke (who assumed command after Cook's death) gifted a selection of every kind of article from the South Seas islands they had visited.[332] This collection had been largely assembled by William Anderson, ship's surgeon aboard the *Resolution*, who had died of tuberculosis in the Bering Sea. Eventually the collection was taken to St Petersburg, Russia, and became part of the Academy of Sciences collection in Kunstkamera in 1780 (Peter the Great Museum of Anthropology and Ethnology). This was the first collection of Cook artefacts from the third voyage to be held in Europe, although none of the fish-hooks are of New Zealand origin.[142]

The Museum für Völkerkunde (Museum of Ethnology) in Vienna, Austria, is one of the most significant ethnological museums in the world, with collections comprising more than 200,000 ethnographic objects. One of its oldest collections derives from the Cook expeditions through the purchase of 238 objects in 1806 at the auction of the Leverian Museum contents in London,[350] although this purchase does not appear to have included any fish-hooks. The Leverian Museum (or Holophusikon) was a private museum of natural history specimens and curiosities that had been accumulated and exhibited from 1775 to 1786 by Ashton Lever,[337, 351] and included the largest collection of Cook artefacts from the third voyage.

Lever offered his museum's collections to the British Museum at a low price, but the offer was refused on the advice of Joseph Banks who stated that there was little of value in them,[345] despite the Pacific material being described as the 'pièce de résistance' of the Museum.[337] Catherine II of Russia also refused to buy the collections, so Lever obtained an Act of Parliament in 1784 to sell them by lottery.[352] The collections were acquired by a James Parkinson who continued to display them after Lever's death in 1788; they were finally dispersed at an auction held in 1806.

The Museum also has an extensive collection of Māori artefacts, including 25 fish-hooks, obtained by the Austrian naturalist Andreas Reischek, who had been selected by Dr Ferdinand

---

[29] National Museum of Ireland AE1893-760
[30] National Museum of Ireland AE1893.761

von Hochsetter to assist in setting up displays at the Canterbury Museum, Christchurch, then under the direction of Dr Julius von Haast. In New Zealand, Reischek's work centred around the Canterbury, Whanganui, and Auckland Museums, but he also collected on his own account, amassing a vast collection of biological specimens as well as many objects of ethnographic interest. Between the years 1877 and 1889 he travelled extensively throughout New Zealand and many of the offshore islands, including the Chatham, Auckland and Campbell Islands, collecting over 15,000 specimens of animals and plants.[142]

Reischek was friendly to many Māori. He visited the King Country with the express permission of King Tawhiao, and on his journey southwards, at Whatiwhatihoe, he received from the King's uncle, Te Witiora, a casket with a huia tail or *hua*. He was thus created a chief and named 'Ihaka Reiheke, Te Kiwi, Rangatira te Auturia'. On later visits he received additional gifts, but he also showed little hesitation in appropriating objects of value without permission, to add to his collections.[352, 353]

FIG.89 **Composite hook, possibly collected during one of Cook's three voyages. Hunterian Museum, Glasgow, E.403/1**

The Hunterian Museum collections at the University of Glasgow, Scotland, were accumulated over four centuries by a number of individuals, but particularly by William Hunter (1718-83) an eminent physician and obstetrician who bequeathed his collection to the University after his death. The collection is a mix of comparative anatomy and pathology specimens, as well as numerous cultural items and natural history specimens. It is the oldest museum in Scotland, as it opened in 1807 and houses material from all three of Cook's voyages (including specimens thought to have been collected by Sydney Parkinson and Joseph Banks). Several Māori fish-hooks attributed to Cook's voyages were originally accessioned as donated by Dr G. Turner (Fig.89), but they cannot be documented as Cook's as there are no items from New Zealand on the 1860 donation list. The shanks of these hooks have been coated in black varnish typical of old Hunterian objects, indicating that they may have been in the collection before 1870. These hooks may be those mentioned by Captain John Laskey in his 1813 account of the museum, in which case they were possibly collected during one of Cook's voyages to the Pacific.[142, 354]

The Pitt Rivers Museum collection at Oxford, England, is regarded by specialists as the most important of the Forster collections and as one of the most important of all the collections made on any of Cook's three voyages, with a total of 186 objects identified as being from those expeditions. The items were acquired by Reinhold Forster and his son Georg during Cook's second voyage of discovery from 1772 to 1775.[355, 356] The collection was selected by the Forsters from a much larger number of objects acquired on the voyage and sent to Oxford in 1776 along with a handwritten *Catalogue of Curiosities*.[294] The Pitt Rivers Museum also holds a collection of artefacts acquired by Joseph Banks on Cook's first voyage and sent to Oxford 18 months after his return, around January 1773.[356]

The Oxford collection has not yet been satisfactorily published, although some individual items have been widely illustrated, and other non-fish-hook items have been studied in great detail. This collection includes approximately 450 Māori fish-hooks collected during the 19th or early 20th century. Among these, less than a dozen were collected prior to the mid-1800s, but many of the hooks do not appear to be of Māori origin. There is evidence that Māori and Polynesian fish-hooks were included amongst anthropological objects transferred from the Ashmolean Museum, Christ College, to the Pitt Rivers Museum in 1886,[142, 355, 356] and they probably originated either from Captain Cook on the second voyage and were donated by Reinhold or Georg Forster, or they originated from two other collections donated to the museum and were obtained by Captain Frederick William Beechey in 1825-28, and Charles A. Pope in 1868-71. Beechey had presented a significant collection to the Ashmolean Museum[31] from his command on the *Blossom*, a northern Pacific surveying voyage.[357] The Pope collection (mostly from North America) from St Louis, Missouri, was probably donated by John O'Fallon Pope (son of Charles A. Pope) who was at Christ Church from 1868 to 1871.[142]

There is not enough distinctive stylistic evidence or concrete documentation to determine whether any of the fish-hooks included in the PRM Cook's catalogue were collected by the Forsters, or if they could even be associated with Cook's voyages.[32] A number of fish-hooks have been assigned Forster numbers[33] but these attributions are tenuous. Catalogue notes, attributed to Assistant Keeper Evans of the Ashmolean Museum, 1884-1908, state: "…it is very plain that all these fish-hooks (No.1281 to 1305) belong to more than one collection and that at

---

[31] Pitt Rivers Museum catalogue notes
[32] PRM Catalogue notes attributed to Peter Gathercole (Department of Anthropology, Otago University, 26 February 1997)
[33] Pitt Rivers Museum #1282, #1292, #1301-1305

some previous time they had been carelessly mixed together. There is not one of Captain Cook's original number labels on any of them, and therefore none may belong to his collection but probably that will never be known now…"

One composite wooden hook with a bone point[34] (Fig.90) has been described as a Māori fish-hook from New Zealand, and was reputedly donated by Joseph Banks.[356, 358] The hook was part of the collection transferred to the Pitt Rivers Museum from Christ Church College, via the University Museum in 1886. This collection comprised a variety of artefacts. Some were originally thought to be from North America, while others were recognised as early Polynesian. Much of this material was from the North American collector Charles A. Pope.[356]

It is unclear how the early Polynesian artefacts came to be mixed among the North American material, and tenuous and circumstantial evidence was used to suggest that rather than being from the Pope collection, the wooden hook was acquired by Joseph Banks during Cook's first voyage, as part of a forgotten collection of Banks' material that had been part of the collection donated in 1773,[356] but had been mislabelled at a later date.[358]

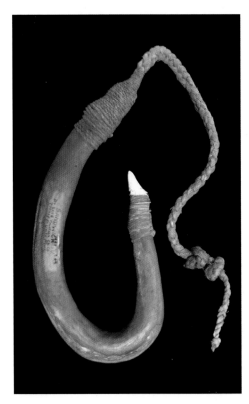

FIG.90 **Composite wood and bone fish-hook from Hawai'i. Date unknown. Pitt Rivers Museum, Oxford, 1887.1.379**

This particular hook is not from New Zealand – the point lashing is typically Hawai'ian, not Māori, it is lashed with unidentified fibre, possibly hibiscus or mulberry fibre (olonga), rather than New Zealand flax, and has old ink writing directly on the wooden shank (partially obscured by the registration number): "Sandwich Ids, Dr. Lee'S Trustees. Ch.Ch., Transf. fm. Unty. Mus."

This hook could not have been included in the collection donated to Christ Church College by Banks in or prior to 1773,[356] as the Sandwich Islands (Hawai'ian Islands) were not visited by Europeans until Cook's third voyage in 1778. Hence, it remains a puzzle how Banks could have acquired a hook that could only have been collected on or after the third voyage. It is likely that this hook is not part of the Banks collection, but came from the collection made by Captain Beechey of the *Blossom* Expedition which was transferred to

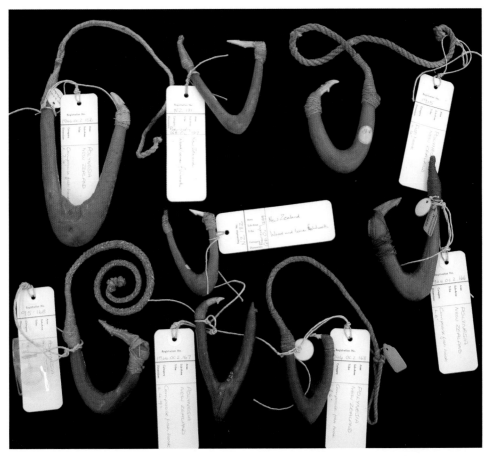

FIG.91 **A selection of composite fish-hooks held in the British Museum, most without dates of collection or exact provenance. British Museum, London. Clockwise from top left: 1944.Oc2.156, NZ181, 1915, 1944.Oc2.166, 1944. Oc2.168, 1944.Oc2.167, centre: NZ 182**

the Pitt Rivers Museum at the same time as Pope's, and was obtained in Hawai'i between 1825 and 1828.[142]

The British Museum in London holds over 3000 Māori objects, including around 350 fish-hooks (Fig.91). The earliest pieces were brought back from Cook's three voyages of discovery during the years 1768-80 and attracted much public interest when they first went on display in the Museum's South Sea Room in 1803. Documentation of that early material was poor and it was not until the end of the 19th century that James Edge-Partington began the task of cataloguing it. In recent years, Adrienne Kaeppler (Smithsonian Institution) has shown that some 28 of the museum's Māori items can be traced to Cook's voyages. The earliest acquisition date for

---

34 Pitt Rivers Museum 1887.1.378

FIG.92 **Pā kahawai from the second Cook voyage.** Humphrey Collection no. 245. Georg-August-Universität Göttingen, Oz 336

any of the fish-hooks is 1875, although many were obtained from earlier collections but lack any provenance details.[294, 332]

The British Museum also received much of the Harry Beasley collection. Beasley was a major ethnographic collector in early 20th century in England whose main interest was in material from the Pacific Islands. In 1928 he opened his own museum, Cranmore Ethnographical Museum, in Chislehurst, Surrey.[283] Beasley wrote a number of journal papers and a book on Pacific fish-hooks in which Māori fish-hooks are particularly well represented.[140]

The Beasley collection is of considerable interest because it includes many unique examples of Māori hooks prepared from materials not found in hooks held in other European or New Zealand collections. For example, amongst the composite hooks are three where the lower jaw of a dog has been used for the point attached to wooden shanks, one made from part of the lower jaw of a horse with a shell point, and others made using cow's horn, pig tusk, and shell – possibly Cook's turban shell. Four lures made from stained moa bone,[35] with bone points (but without pāua shell inlays) and intact shank-lines are unique, and other similar lures are not known in any collection examined by the author in Europe or New Zealand, with the exception of two lures in Museum of New Zealand Te Papa Tongarewa collections.[36] The unusual nature of many of the hooks in the Beasley collection suggests that they may have been made as replicas or fakes, possibly commissioned by dealers selling hooks to Beasley. Although the use of dog jaw points has been well documented in archaeological sites,[201] no examples are known from any of the Cook voyage material.

---

[35] British Museum BM 95-408
[36] Museum of New Zealand Te Papa Tongarewa ME00227

The National Museum of Scotland, Edinburgh, is a notable exception from other museums in that it has continually supplemented the collections with touristic material from New Zealand from the late 19th century and throughout the 20th century.[303] [306, 307]

Ethnological souvenirs have an ambivalent status in museums. Where it is known that they were produced for the tourist market they are frequently regarded as less valuable than other similar items manufactured for native use. Tourist souvenirs are often seen as examples of the 'invention of tradition' or non-indigenous material, and are therefore non-authentic, and rarely knowingly included in exhibitions. This collection is important as it allows comparison of objects known to be made as souvenirs with pre-European contact examples.[142]

Among the collection, which includes 41 Māori fish-hooks, are several wood-backed pā kahawai with pāua shell lining. While most of these lures are undated, they are typical examples of pā kahawai that were made in large numbers in the latter part of the 19th and early 20th centuries.[2] One example[37] among a number of items collected in 1846-50 at the Bay of Islands by Dr John Thomson, fleet surgeon on the *Rattlesnake* Expedition, represents the earliest documented example of a wood-backed lure or pā kahawai with pāua shell lining (National Museum of Scotland, Edinburgh, A06256). Prior to this, all known lures were made with a simple wood shank (Fig.92).

Material from Cook's voyages held in the Berne Historical Museum, Switzerland, was donated by John Webber, artist on the third voyage, and represents the largest extant, well documented, third-voyage collection. Unlike the collections made by Cook, which often comprised ceremonial artefacts or gifts, the Webber collection is of more typically ordinary, useful things.[294] Most of the items are from Hawai'i, Tonga and the Society Islands, with only a few from New Zealand. Although a Māori fish-hook was included in the shipping list of objects sent to Berne,[359] none is present in the collection or mentioned in earlier lists.

The collection of ethnographic objects from Cook's voyages held at the Museo Zoologico e di Storia Naturale della Specola, in Florence, Italy, was the first Cook collection to be described and published.[294, 360] There is no documentation of how the objects were brought to Florence, and the evidence that they are from Cook's voyages is circumstantial, although at least some may be from the sale of the Leverian Museum in London in 1806, and others may have been purchased in 1779.[294] In 1893 Enrico Giglioli described a number of artefacts in the museum which he stated originated from Cook's visit to Queen Charlotte Sound (New Zealand), but it is not possible to confirm that the objects came from Cook's voyages, let alone to locate them to a particular site.[294] Two fish-hooks are included among the items described by Giglioli. The first is described as incomplete, being only the bone point of a hook which Giglioli attributed to the

---

[37] National Museum of Scotland V.2007.300

point of a wood shank lure lined with pāua shell (pā kahawai) without citing any evidence. No wood and pāua shell pā kahawai lures are known from Cook's voyages or were collected by other early European explorers. The second hook described by Giglioli comprises an almost complete composite hook with a flax snood lashing, but lacking a bone point.

The Cook voyage objects, gathered during a period of imperial expansion, endure to the present day and represent the earliest exchanges between Māori and European culture, and are traditional taonga that have not been influenced by the impact of European culture and technology. Their perceived importance as unique specimens of pre-contact Māori craftsmanship has increased considerably in recent years and is the focus of much museum-based research. [303]

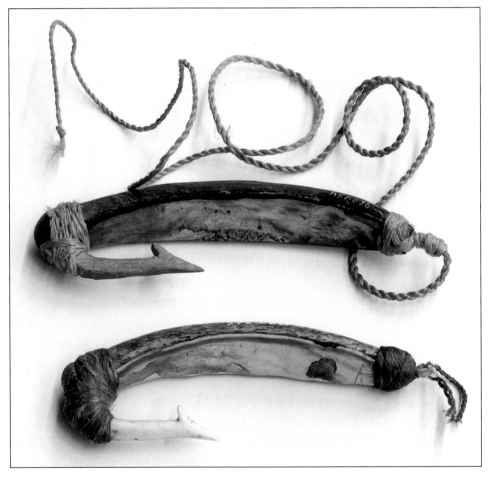

FIG.93 **Prices for collectable hooks, especially pā kahawai with inlaid pāua shell, have risen significantly in recent years, but not to the same levels as hooks with suggested 'Cook voyage' provenance. Puke Ariki, New Plymouth, A80_246, A57_876**

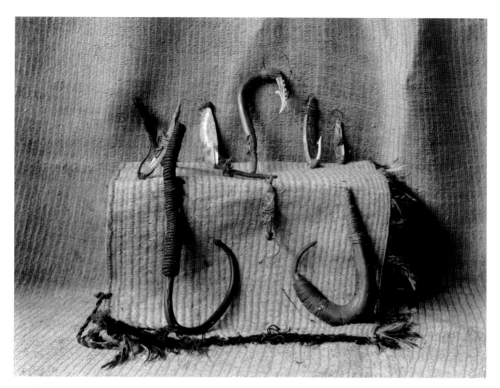

FIG.94 **A display of a selection of hooks and lures made using both traditional materials and metal components, photographed between 1890-1930. Alexander Turnbull Library, Wellington, New Zealand, Ref 1/1-007560-G**

As well as museum collections, many examples of traditional artefacts are held in private collections and have become highly collectable, not only for their perceived value as tribal art and cultural objects,[278] but as a financial investment in uncertain economic times. As early as the late 1800s, many exhibitions and curiosity shops featured attractive displays of Māori fishing equipment for sale to tourists and collectors (Fig.94) and these items have been traded and auctioned between collectors since that time.

Auction prices for high quality 18th and 19th century Māori pieces have soared exponentially in the early years of the 21st century, particularly following the success of several international touring exhibitions highlighting Māori culture (Fig.93). At a premium are examples in good condition with good provenance. These may reach prices of over 100 times those achieved in the late 20th century.[281, 285, 361] As values rise there is an incentive for unscrupulous vendors to increase the number of copies, replicas and fakes; therefore, provenance becomes increasingly important.

9

Taiao: *fisheries conservation*

"I write this to express my surprise that men endowed with reason can think, in this nineteenth century, that five hundred bundles, or even five million bundles of fish can have any effect on the numbers of fish in the sea. Nature is so prolific that the more we catch the faster they multiply... It is all bosh! There is selfishness at the back of it..."

Fishmonger A. Sanford, letter to *New Zealand Herald*, November 1886 (Hector 1897: 18)

Possible impacts on coastal fish stocks as a result of harvesting in pre-European contact New Zealand have been widely debated.[45, 47] In most regions relatively low population densities of Māori had little or no effect on the profusion of fish stocks, particularly the vast schools of pelagic species such as barracouta in southern waters, and kahawai and mackerel (kōheru, *Decapterus koheru*; tawatawa, *Scomber australisicus*; hāture, *Trachurus* spp.) further north.

Archaeological studies reveal some very localised impacts on fish size, but indicate little significant effect on the abundance of fish species over time.[186] Evidence of the impacts of harvesting on fish populations could be obscured by gradual climate change or by changing patterns of fishing technology targeting different fish stocks and habitats as fishing pressure increased.[47, 362-364]

Changes in sea temperature affecting fish growth rates and distributions is the probable cause of changes in the mean size of barracouta, and the abundance of snapper represented in South Island prehistoric midden sites over time.[47, 364, 365] Barracouta is a pelagic species found in immense numbers offshore, while snapper is found in warm temperate waters and is less prolific in cooler southern waters. Non-migratory coastal reef-dwelling fishes are more vulnerable than schools of pelagic species offshore, and evidence suggests some localised overfishing.

At several archaeological sites an increase in mean fish size over time is apparent for resident species including blue cod, snapper, scarlet wrasse, spotty, and banded wrasse. This may be an indirect result of Māori fishing pressure, representing changes in fishing techniques.[44, 47, 366] This increase in mean fish size over time challenges the widely held view that fishing pressure results in a lowering of mean size and suggests a hypothesis that if people are permitted to take as many small specimens as they wish, sustained fishing of a species sensitive to human predation may lead to an increase in mean size.[44]

However, this concept that a take-everything approach may not be as damaging to coastal ecology as is widely believed is simplistic, and it is more likely that the trend in increasing sizes in archaeological sites represents a change from early inshore fishing with nets and traps (targeting smaller juvenile fishes), to fishing for larger fish with baited hook-and-line on deeper reefs as the more accessible reefs were depleted. Line fishing would have selected larger individual fish and result in an apparent, but misleading, increasing size trend in midden material.[47]

The abundance of fish stocks in shallow coastal waters that were available to pre-European Māori made catching adequate numbers for daily food requirements relatively easy using nets, traps, spears, and hook-and-line, or even simply gathering fish such as grey mullet that accidentally leapt into a canoe, as was a common occurrence in northern harbours such as the Kaipara.[146]

The Europeans introduced new preservation techniques using ice and canning which enabled more fish to be taken and processed than was required by the local population, thus the demand began to exceed the productivity of the fish stocks and the fisheries expanded to wider areas. The undocumented seasonal fishing activities and unrecognised conservation techniques of Māori were lost, and some fish stocks began collapsing even before the end of the 19th century.

Fisheries conservation and management in the 19th century

By the mid-1800s the increasing European population in New Zealand had exceeded that of Māori.[367] The abundance of fish was obvious, and commercial fishing became an area of interest as a potential new industry for the settlers, but without consideration of how traditional fisheries that underpinned Māori culture would be affected.[146]

At a meeting of the New Zealand Advancement Society in Auckland as early as 1848, it was suggested that fish cured for foreign markets was the colony's "…most available export…" and that the numerous large fish around the coast would amply repay any investment.[38]

In 1871, the Hon. John Munro, Member of the House of Representatives, wrote a memorandum to the Joint Committee on Colonial Industries in which he stated:

"I do not only believe, but I know, that there is an inexhaustible source of national wealth swarming unmolested round these islands, and on sunken rocks not yet discovered, that

---

[38] *Nelson Examiner* and *New Zealand Chronicle*, 18 November 1848: 149

will yet be a profitable resource for the laborious fisherman, and contribute largely to the aggregate prosperity of the country…"[88]

Europeans began to expand their interests, at first into the lucrative oyster trade, then into other fisheries, and Māori involvement in commercial fishing began to decline. In 1868 an editorial in the *Otago Witness* noted that:

"…it is to be regretted that so little attention has yet been given to the prosecution of the fisheries… attention has hitherto been so much engrossed on sheep farming and agricultural operations and gold mining, that little has been done to reap the rich harvest of the sea… we cannot feel but assured that the Colonial fisheries will not be long overlooked… the riches of the sea will eventually engage as anxious a consideration as the riches of the land… it needs but a few boats' crews to make a systematic beginning to prove that the large amount of wealth at present swimming unheeded on their coasts… with the rapid progress of settlement, fishing and the curing of fish would soon become a profitable occupation… every bay and inlet of New Zealand abounds with fish…"[39]

In 1872 W.H. Pearson, Commissioner of Crown Lands at Invercargill, promoted Stewart Island as a settlement for immigrants from Orkney, Shetland and the Western Islands of Scotland, stating: "…the fisheries around Stewart's Island promise, not only comfortable subsistence, but wealth… there is every evidence that the supply of fish, most of very superior quality, is inexhaustible… the sea appears to literally swarm with them…"[368]

At the time the new government had little information available regarding the fisheries of the colony. A Fisheries Commission had been appointed in 1868-69 to gather basic information, such as what fish were common in New Zealand; however, all that had been learned was what fishermen and fishmongers could relate, as there was little information to be found.[88]

In 1875 the government published *The Official Handbook of New Zealand* which set out to describe the opportunities for British colonists and to encourage business investment. Mr Seymour, Chairman of Committees of the House of Representatives, and Mr Rolleston, Superintendent of Canterbury, noted that:

"The coast appears to teem with useful and excellent fish, and a further extension of this industry is to be expected… Amongst the industries which might be carried on to advantage in addition to those already present in operation… may be mentioned fish-

---

39 *Otago Witness*, 18th January 1868: 14

166   Te Matau a Māui

curing… The Pelorus and Queen Charlotte Sounds would form admirable fish-curing stations on a large scale… Fish of all kinds, and oysters, are plentiful, and the herring fishery offers every inducement for a profitable investment… At present, the industry is not prosecuted to any great extent… The culture of oyster-beds would also be found profitable, and capable of great extension…"[367]

Government interest in fisheries was initially limited to preservation and propagation of the newly introduced freshwater trout and salmon. Legislation was introduced in The Fish Protection Act 1877 primarily to protect trout, prohibiting the use of dynamite or other explosives for catching fish in rivers and streams. An 1878 amendment, The Fisheries (Dynamite) Act, was required to expand protection to fish in "lakes" which had been omitted from the definition of waterways in the original Act. The amendment extended protection to "…all fish ordinarily inhabiting the waters of the Colony, whether fresh or salt water…"

In 1886 the government introduced legislation to establish a closed season for mullet in northern waters. This was one of the first direct government interventions into the conservation of sea-fisheries in New Zealand and was immediately decried by the public and debated widely in local newspapers. General opinion was that the impact of harvesting on the vast schools of fishes surrounding the coasts was negligible. Many early European explorers in the 18th century, followed by missionaries and settlers in the early 1800s, had published accounts of their travels, providing numerous anecdotal comments on the local abundance and ready availability of fish throughout New Zealand.[23, 59, 76, 89, 255] The apparently endless supply of fish was widely regarded as a potentially lucrative fishing industry for the developing colony.[74, 88, 369]

In 1897 the Member for Ashburton, the Hon. Mr Wright, introduced a Bill into parliament making it illegal to take or sell flounders from Ellesmere that were less than 11 inches (27.9 cm) in length, following concerns that the estuarine lake had been overfished. It was noted that the cause of the depletion was unknown, and newspaper correspondents claimed that the number of natural enemies of the small fish was so great that defining restrictions for a minimum size was immaterial.[40]

Even those who recognised early signs of overfishing remained optimistic: in 1913 the New Zealand Fisheries Commissioner, Lake Falconer Ayson noted:

"…some of the old fishing grounds within a certain distance of the larger centres are not now producing anything like the quantity of fish which they have done formerly, and in several places fishermen find it necessary to keep moving further afield in order to

---

40 *Otago Daily Times*, 12th November 1897: 4

get the supplies required. The cause of this decline is, I consider, due to overfishing and the predominance of sharks, dogfish and other enemies of our market fish… the areas I have mentioned as suffering from overfishing are not very extensive; in fact they may be considered as a mere bagatelle in comparison to the fishing grounds round our coasts which have as yet not been exploited…"[370]

Until the beginning of the 21st century, extinction of a marine fish species had been considered at best unlikely, if at all. In an inaugural address to the 1883 International Fisheries Exhibition in London, prominent biologist and philosopher Thomas Huxley asserted that overfishing or "permanent exhaustion of the sea fisheries" was scientifically impossible. Huxley stated:

"…probably all the great sea fisheries are inexhaustible… the multitude of these fishes is so inconceivably great that the number we catch is relatively insignificant… the great shoals are attended by hosts of dog-fish, pollack, cetaceans and birds, which prey upon them day and night, and cause a destruction infinitely greater than that which can be effected by the imperfect and intermittent operations of man… Any tendency to overfishing will meet with its natural check in the diminution of supply… this check will always come into operation long before anything like permanent exhaustion has occurred…"[371]

This view was probably shared by most fishers at the time, and expressed in letters to the *New Zealand Herald* in 1897 during debate surrounding the Northland mullet fishery:

"The wheel of nature is always turning, assuming different forms, never lessening the whole one atom, but so regulated by Him that fallen man is powerless to control or affect in the least…"[41]

"In a thousand years there has never been any fear that the enormous takes of spawning fishes should have any effect on the schools… These facts apply to all our fishes in New Zealand, so that we need not fear that in a few years there will not be any fish, or that any particular kind of fish will be destroyed by overcatching… [we] will not be frightened by the silly remarks of some of your correspondents…"[42]

Although Huxley later changed his opinion, for most of the 19th century, and until almost

---

[41] A. Sanford, letter to *New Zealand Herald* 15th November 1886, *in* Hector 1897
[42] C. Bishop, Fishmonger, letter to *New Zealand Herald* 13th February 1896, *in* Hector 1897

FIG.95 **Processing mullet in a Northland factory, circa 1920. Alexander Turnbull Library, Wellington, New Zealand, PAColl 3077**

the last few decades of the 20th century, this attitude persisted – that the sea was limitless and that there would always be plenty more fish in the sea.[2, 146]

Protection of mullet

In the early 1800s the Kaipara Harbour supported a population of between 2000 to 3000 Māori, with the main food supply based on the huge schools of grey mullet (and possibly sand mullet, *Myxus elongatus*), both in the harbour and on the outer coast.[146] Around 1840 Kaipara was one of the busiest ports in the country and Māori supplied dried fish for sale to the trading vessels.

During summer months "swarms" of mullet were reported throughout Northland and in 1854, Māori supplied 27 tons of mullet and other fish to the Auckland market over a period of six months.[65] The mullet resembled those of the "old country" but were described as being far superior in flavour and the Europeans began smoking and canning them for local markets and export to the Australian goldfields. By the early 1880s several commercial canning factories had been established in the Kaipara and in other areas, including Hokianga and Bay of Islands (Fig.95).[146]

FIG.96 Refrigeration enabled fish to be transported and stored. Townsend and Paul's fishmarket, Wellington , circa 1910. Alexander Turnbull Library, Wellington, New Zealand, ID 1/2-047903-G

Initially the European canning factories were dependent upon Māori expertise for their supplies.[370, 372] Each factory processed between 95 and 120 tons of mullet per year;[372] however, the catch soon declined. Following requests from fishermen, in 1886 the government established a closed season throughout Northland from 20 December to 1 February annually, to protect the fish stock during the presumed spawning period, but two years later the protected area was reduced to cover only the inner Kaipara Harbour. Despite this, many local fishermen and members of the public were sceptical about the need for protective measures. In 1886, an editorial in the *New Zealand Herald* proclaimed:

"…Friday last we published a telegram from Wellington stating that, as representations had been made to government on the subject, it had been determined to have a close season for mullet… strong evidence had been furnished that these valuable fish do not now appear on our coast in anything like their former abundance… we are further told that the government had taken much trouble to 'procure all available evidence as to the increasing scarcity of fish'… as to the mullet becoming scarce from the fishing that has been carried on, we simply do not credit the statement. If the few people now in New Zealand have already decreased the number of fish on our shores, when the colony is thickly populated we shall not have a fish left. The assertion seems to us ridiculous…"[43]

The mullet fishery continued, but within a decade competition between rival canneries and fishermen resulted in claims of a further decline in the fishery and calls for an extension of the closed season. Sir James Hector, Director of the Geological Survey and Colonial Museum, was asked by government to examine the fishery in 1895-96 and report on the need or otherwise, for a closed season.[372] The lack of baseline information on the biology of mullet, and even the correct identification of the species taken, limited the conclusions that could then be drawn. A proposal to extend the closed season from 1 December to 31 March, and expand the protected area to the entire Kaipara Harbour was gazetted on 9 September 1895, then revoked two months later (before it came into effect), on advice from Hector amid protests both for and against the closure from local fishermen.

In November 1896 Hector's final report was presented to government. Hector asserted that no closed season was required but he recommended further investigations be carried out.[372] The government, taking note of Hector's comments recommending other research, directed resources into experimental trawling and artificial propagation of foreign (European and North American) species to enhance the country's fisheries, and the mullet fishery was largely ignored over subsequent decades. The government removed all protection for mullet and landings fell dramatically, with only 45 tons being landed in 1897. Once government export subsidies were removed in 1905, the factories closed and landings failed to exceed 8 to 9 tons per year until the fish stock partially recovered in the late 1930s.[146]

Fisheries development

In the 1880s the introduction of refrigeration technology (Fig.96) enabled New Zealand to establish a lucrative export industry of meat and dairy products, and the country became a grasslands-based economy.[373] Trial consignments of frozen fish were sent to Europe as early as 1883 when the *Otago Witness* reported the ship *Dunedin* sailed on 20 January with one of the first shipments of fish for "scientific purposes" consisting of four trout around 9 lb each in weight, and a few groper and barracouta in a refrigerated chamber.[88]

Later that year the New Zealand Shipping Company vessel *Mataura* landed another consignment of fish in Britain. Several dozen frozen flounder and mullet were sold at London's Central Fish Market and it was noted that the fish were in better condition than many from other English ports. But the quantities of exported frozen fish remained small, with less than a ton exported in 1886,[65] while in 1893 a shipment of moki failed to sell in London.[44]

In 1885 the Hon. J. McKenzie presented a paper to government on the development of

---

[43] Editorial, *New Zealand Herald* 26th October 1886

[44] *Akaroa Mail and Banks Peninsula Advertiser*, 27th January 1893: 2

colonial industries and extolled the abundance of fish.[374] He described his observations in a letter to the Minister of Finance, Sir Julius Vogel:

"...I examined the coast northwards as far as Whangarei; found schnapper [sic], mullet, kahawai, and bream of fine quality; but as the weather was so bad, I did not devote much attention to this locality, further than to satisfy myself that fish in countless millions frequent the neighbourhood of Great and Little Barrier Isles, and the Firth of Thames... the whole [Kaipara] harbour from Helensville, at one end, to Aratapu, at the other end, a distance of over 80 miles, seemed to be actually swarming with the largest and finest mullet in the world... Although I examined the coast-line from Kaipara Heads to Waitara, near New Plymouth, including the harbour or bays of Manukau, Waikato, Whangaroa, Aotea, and Kawhia, I am not in a position to say that fish always inhabit this region, but I am satisfied that large shoals of schnapper [sic], mullet, and kahawai are to be found here, during some portion of the year in-shore, and most likely off-shore all the year round. I found soles or flounders, kelp-fish, mullet and bream everywhere in-shore, and also many varieties of small but very excellent fish that I cannot name or classify. After rounding Cape Egmont, rough weather prevented me from fishing until off the island of Kapiti, and off that island and the island of Mana, I hooked groper, and netted moki and rock-cod, and got also cray-fish, kelp-fish and butter-fish... I minutely fished all of the Picton Sounds, and also tried the off-shore or deep-sea fishing off Cape Campbell. I found the whole of this region actually alive with fish... The most abundant fish is the Picton herring... it is found here in immense quantities, it could be so cheaply procured that the export trade in this article alone should rival in a few years the herring trade from the North of Scotland... I did not get a chance to try the fishing on the [West] coast, owing to rough weather until close off Martin Bay, and here I commenced to meet with fish in such numerous shoals that from here to off Oamaru, inshore and offshore, I believe millions of tons of fish could be caught yearly. It is simply a question of proper appliances, and finding out the best and quickest modes of catching the fish; for the fish are there in countless millions, and natural harbours abound from Milford Sound to Oamaru. The Sound swarms with blue-cod, moki, trumpeter, rock-cod, and cray-fish; and offshore there are ling in great quantities, and also groper... Ruapuke Isles, off Bluff harbour, in Foveaux Strait, swarms with moki and trumpeter... Chaslands Mistake, on the mainland, commands splendid moki fishing grounds, and also blue-cod, rock-cod, and trumpeter fishing, and here I began to meet with the barracouda in large numbers, and found them all the way northwards to off Oamaru; but off Cape Saunders and Otago Heads seems to be a central gathering-ground for countless millions of those fish for several months in the year… Ling and groper, in great quantities, I found from off Chaslands Mistake to off Timaru. Those fish are found sometimes in-shore, but to get them in quantity they must be fished for off-shore. Otago Harbour commands most extensive and valuable barracouda, groper, ling, rock-cod, and cray-fish fishing, and with proper fishing-smacks Otago Harbour could also command the blue-cod fishing. The kinds of fishes that I have satisfied myself can be obtained in large quantities cheaply, and fit for export trade, tinned, wet and dry salted, and smoke-dried, are Picton herring, in Cook Strait. Blue and rock-cod, moki, trumpeter, groper, ling, barracouda, cray-fish, cockles, flounders, trevalli, silver-fish, mullet, kelp-fish, gurnard, and about twenty other varieties, including a kind of mackerel, abound on the coast of both islands, and tinning and curing factories would use all in their season if ever established. But the other kinds I have mentioned, along with schnapper [sic] and large mullet of the North Island, are the kinds to make the trade with; and no other country in the world has such a variety, and distributed around its coast so well…"[375]

McKenzie was supported by R.A.A. Sherrin, Commissioner of Trade and Customs, who felt that the fisheries of New Zealand would "…surpass the known mineral wealth in total value as well as in permanency and stability…", and he described the fishing resources of the colony reported by the network of lighthouse keepers around the coast in glowing terms. Sherrin believed millions of tons of fish could be caught yearly. It was simply a question of proper appliances, and finding out the best and quickest modes of catching the fish; for the fish were there in "…un-ending abundance…"[88] He also noted that the fisheries would not develop themselves and the absence of any government gathering of statistics about which species were caught, on what grounds, and in which seasons, or promotion of scientific study to learn the habits, food and spawning of the various kinds of fishes, resulted in a complete lack of information, from which no industry could develop.

In a review of the 130 fish species known from New Zealand waters at that time, Sherrin compiled the scattered information into a *Handbook of the Fishes of New Zealand* (1886), which included details about each of the species, and the industry including fish culture methods, as well as canning and smoking procedures, both in New Zealand and overseas. He also included observations on the sealing and whaling industries, which he felt had almost been destroyed by "free trade" before their histories were written.[88] Sherrin described fisheries for each species in some detail. For example, the pilchard (mohimohi, *Sardinops neopilchardus*) was described as:

"…a true representative of the herring kind in these seas and it visits the east coast of Otago every year in February and March, and when the schools migrate they extend as far as the eye can reach… so densely packed are they some years that by dipping a pitcher in the sea it would contain half fish, so that if large boats and suitable nets were employed thousands of tons could be caught… in Queen Charlotte Sound shoals were observed throughout the year… hauls made average 1½ to 2 tons, but at times 10 tons have been landed… they are very abundant in the Auckland waters… especially so at the Thames… There are large shoals of the pilchard at the head of Milford Sound… tons of them are thrown onto the beach in Freshwater Basin…"[88]

The demand for pilchards (sold as "Picton bloater") was high, but the product had to compete with kippered or pickled herring from Europe, and the ready availability of other fresh fish from local Māori suppliers prevented a viable industry from being established. Although good fishing grounds for pilchards and many other species could be found around the coasts of New Zealand, the problem was the lack of a market. Local boats caught and sold fresh fish directly off the wharf, but the fishing was generally on a small scale, and the colonists could easily catch fish for themselves.

In 1888, Peter William Barlow, an early settler in the Kaipara District in northern New Zealand, commented on the potential of commercial fishing, noting that because of the abundance of fish, all that was required was capital to develop the industry.[376] Barlow related tales of pataki or flatfish that could be caught in the Kaipara Harbour with a net in "boatloads", and of fishing for snapper (using lines with two hooks), catching fish at the rate of 60 to 70 per hour, each with an average weight of about 9 lb (4.5 kg).

In 1892 a surveyor, Robert Paulin, commented on the size and abundance of "trevalli" (blue warehou, *Seriolella brama*)[156] at Jackson Bay in South Westland. He described how 14 lb fish (~6 kg) could be caught at a rate of four dozen in a few hours, and when salted, sold at the nearest township, Hokitika, several hundred kilometres north.[377] At the peak of the West Coast goldrush in the 1860s the population of Hokitika reached 6500 (with many thousands more in the district nearby); however, by the end of the century the population had declined to around 2000.[378]

A 'Fishing and Fish Procuring Association' had been set up at Jackson Bay in 1875 and large numbers of fish, including moki, blue cod, and groper were shipped to Hokitika where the fish were described as being "…of esculent variety and quantity sufficient to startle and baffle any minute description even from the hands of an edacious reporter…"[45] More fish could be caught than could be processed and cured: it was reported that 15-16 dozen moki could be caught in a day, and on one occasion, 24 large groper were caught in four hours using two handlines.[46] However, the cost of shipping freight on the small coastal steamers made the enterprise uneconomical.

FIG.97 **Fishermen loading their catch onto a truck. Island Bay, Wellington, 1930s. Photograph by Sydney Charles Smith. Alexander Turnbull Library, Wellington, New Zealand, 1/2-04 7912-G**

The lack of local markets and poor transportation systems to reach population centres (Fig.97) prevented any large-scale commercial fishery from developing until some other means of preserving the fish – such as refrigeration – was readily available. Refrigeration enabled unprocessed fish to be stored, and transported to markets via the new railways, but once there they faced direct competition with other more desirable species and fresh rather than frozen fish. It can be inferred from Fisheries Inspectors' statements in the Annual Fisheries Reports during the period 1906-23 that the fishing industry was limited by market demand, with more desirable fish species being landed directly into the cities – the main market after export subsidies ceased.

Trawling and acclimatisation of new species
The government recognised a need to assist the developing fishing industry, and exploratory trawling expeditions in the early 1900s were the first to attempt to quantify New Zealand fish

---

[45] *West Coast Times*, 21st September 1875: 2
[46] *West Coast Times*, 18th January 1876: 2

stocks. Although by this time some concerns had been expressed over some localised fisheries, and some decrease in hāpuka, snapper and blue cod stocks had been noted in a few restricted areas, it was generally considered that the sea was not fully exploited and conservation was of little, if any, concern.

Initial trawling by the fishing vessel *Redcliff* off Otago in 1868 had shown that trawling could be more efficient than lining,[65] but it was the invention of otter-boards to hold the trawl net open that revolutionised this fishing method and a new design was patented in 1894.[244] Large steam vessels fitted with trawls had begun experimental fishing around New Zealand in the 1880s, and with government subsidies from about 1895, vessels were soon operating and landing catches into the coastal towns of Auckland, Canterbury and Otago.[65] Between 1900 and 1907 the Marine Department invested in exploratory trawling with a series of surveys by the steam trawlers *Doto*, *Nora Niven* and *Hinemoa*.[338] Unlike trawlers powered only by sail, the new steam trawlers could supply fresh or frozen fish in all but the worst weather, but were limited to known waters by the absence of good bathymetric charts to show reefs and foul ground.

The British Royal Navy vessel *Acheron* had carried out the first systematic hydrographic survey in 1848-51, but soundings were limited and it was not until the early 1920s that effective echo-sounders enabled more accurate charts to be developed.[379, 380] Accurate charts were essential to enable the trawlers to avoid rough ground and reefs which could then be targeted by longline fishermen (Fig.98).

In the 1860s trout and salmon had been successfully introduced into New Zealand and the prevailing consensus was that equal accomplishments could be achieved in introducing the finest marine food-fishes of Britain – cod, turbot, and herring – to create fisheries that would rank among the most valuable assets of the colony.[372, 381] In the late 19th century, considerable success with artificial propagation of fish stocks had been achieved in both the United States and Europe. James Hector referred the government to the success of these operations. He submitted a paper by a Dunedin scientist and schoolmaster, G.M. Thomson,[381] that promoted development of artificial propagation as a means of

FIG.98 **A 234-pound bass caught in Cook Strait, Wellington, February 1950.** (*Evening Post* Wellington). Alexander Turnbull Library, Wellington, New Zealand, 114/117/13-G

FIG.99 **Portobello Marine Laboratory on Otago Peninsula was built in the early 1900s as a fish hatchery.** Photograph circa 1910. Hocken Library, Dunedin

enriching the sea fisheries, stating: "…the only measures to be recommended for the conservation of mullet and other fishes that spawn in the sea is artificial propagation…"[372]

Acting on this advice, in 1904 the government directed fisheries research and resources into establishing a fish hatchery and marine investigation station at Portobello on the Otago Peninsula (Fig.99), to achieve the goals of acclimatising new species to New Zealand. Attempts in 1886 to import herring ova had failed, even though the eggs had been kept in iced water to arrest their development during the long sea voyage. Thomson considered such attempts doomed without a fish hatchery ready to receive live eggs,[381] and he worked hard at promoting the idea of a marine station to scientific and commercial circles through the New Zealand Institute (known as the Royal Society after 1933) which had been established in 1851, and whose objectives included developing the resources and capabilities of the colony.

The main role of the Portobello hatchery was to help establish European species of fish, crustacea and shellfish – such as lobster, edible crab, turbot, herring and others – in New Zealand. Although millions of young of many species were released into Otago waters over those early years, none is known to have survived,[382] and the marine fish hatchery did not achieve the success hoped for. Today finfish aquaculture has become an established industry, so while

FIG.100 The *Florence Kennedy II*, a popular Auckland charter boat, returns to port in 1958 with over 3000 snapper caught by 48 anglers in a 4-hour fishing trip in the Hauraki Gulf. © New Zealand Herald.

Thomson's and Hector's championing of hatcheries for the purpose of introducing European fish species may have been inappropriate, from a more general perspective they were certainly ahead of their time.[146]

In 1937 a report by a special Government Sea Fisheries Investigation Committee set out to review the condition and prospect of the sea-fishing industry of New Zealand, including investigations into any matter relating to the exploitation and conservation of the sea fisheries. The investigation coincided with the 1930s Depression when prices were poor and fish landings were low.[383] Following the report, licensing restrictions were established in response to concerns of localised overfishing (this was done away with by 1963). The committee made 200 recommendations and concluded that the best interests of New Zealand would be served if the fish-export trade were almost totally dispensed with and recommended that policy concentrate on an increased distribution of fish to local markets throughout New Zealand.

Fisheries decline

English common law, focused on a common property belief, led to a largely unregulated open access approach to fisheries management. Under open access regulation, fishermen had strong incentives to catch all the fish they could (at least until the cost of catching the fish exceeded

FIG.101 **Ten big billfishes caught in one day by Captain Mitchell (left) and Zane Grey, 15 April 1926.** Sir George Grey Special Collections, Auckland Libraries, AWNS-19260415-40-1

the market price), leading to overfishing.[384, 385] Combined with government subsidies which enabled fishing to continue long after costs exceeded any profits, the outcome of open access was inevitably commercial extinction of some fisheries.

Fish stocks remained high around many parts of the New Zealand coast into the 20th century. In 1903 the *Marlborough Express* reported that after six or seven hours' fishing, a group of recreational fishermen in the Sounds had landed two and a half tons of groper, butterfish, rock cod (*Helicolenus percoides*?), moki, teraki [sic] (tarakihi) and other kinds.[47] However, by October 2008 the decline of fish stocks within the Sounds had reached such a level that the New Zealand Ministry of Fisheries banned recreational fishing of blue cod in all enclosed waters for a period of three years. This was followed by a re-opening of a limited season in 2011, with a two-fish bag limit, as well as minimum and maximum size limits. Commercial fishers in the area also agreed not to fish in the enclosed Sounds while the closure was in force, in an effort to allow stocks to rebuild.

In 1862 local observers in Dunedin reported that "…In the harbour are caught flounders, mullet, trumpeter, gar fish, codling, silver fish, trevelhas [sic] etc., all of which are much scarcer

---

[47] *Marlborough Express*, 4th November 1903: 4

than they used to be. Indeed, unless some sort of close time be allowed, these fishes will be known in Otago harbour only in name... the nets are worked nearly every tide and as the fish have got scarce, smaller ones are caught and sent to market..."[48] Similar concerns regarding the taking of small fish were expressed in 1876-77,[386, 387] and discussed again in local papers in 1886 when it was suggested that dredging and other harbour works were as much to blame as overfishing.[49]

Regulations limiting the capture of small fish were introduced in January 1888, but were not rigidly enforced, although at least one fishmonger was fined £1 for selling two undersized "trevalli".[50] Several scientific papers on the fishes of Otago Harbour published between 1906 and 1938 made no mention of any decline in fish stocks, referring only to the "super abundance" of small pelagic species.[388-392] On 1 October 2010, the Ministry of Fisheries announced that new fishing regulations were being introduced for a taiapure in the East Otago Harbour to ensure long-term sustainability of the fishery and rebuild stocks. The regulations included reduced recreational bag limits for certain species, and a maximum limit of 10 for all finfish.[393]

In the early 20th century, weekly recreational fishing trips on charter vessels were becoming increasingly popular with the New Zealand public and numerous local clubs were established (Fig.100). It was reported in 1914 that off Akaroa, Banks Peninsula, launch parties could expect to catch about five hundredweight (~250 kg) of fish per day, including groper up to 25 kg,[51] and in 1924 the *New Zealand Herald* reported that a party of fishermen aboard a fishing yacht caught two tons of groper in one weekend off Mayor Island in the Bay of Plenty,[52] while individuals could easily catch up to 80 fish per trip.[53]

Recreational fishing, and particularly big game fishing off the northern coast, was popularised internationally by Zane Grey with the publication of *Tales of the Angler's El Dorado, New Zealand* (1926) which described catches of six to ten or more sharks, swordfish and marlin up to 975 lb (440 kg) per day (Fig.101).[394, 395]

David Graham, in his book *A Treasury of New Zealand Fishes* (1956)[156, 396] described how in the early 1900s it was not unusual for daily catches of ling off Otago Peninsula and other coastal waters to reach three or four dozen per day, while in 1900-05, fishermen could hook two to three dozen groper per hour off Otago Peninsula, and between 1922 to 1927 two men working could catch 5 to 15 dozen large fish up to 80 lb (36 kg) per day. By 1930-33 observers noted that ling were no longer available and it was possible to fish all day and not catch one, or perhaps just one or two. After the introduction of longline fishing in the 1930s, with boats using up to 180 hooks on as many as four lines each day, the groper fishery declined markedly and Graham reported

[48] *New Zealand Spectator and Cook's Strait Guardian*, 23rd July 1864: 3   [49] *Otago Daily Times*, 26th October 1886: 3

[50] *Star*, 22nd October 1888: 3   [51] *Akaroa Mail and Banks Peninsula Advertiser*, 27th February 1914: 2

[52] *Evening Post*, 10th January 1924: 4   [53] *Auckland Star*, 18th November 1901: 4

FIG.102 **A catch of groper, circa 1910. Photographer unidentified. Alexander Turnbull Library, Wellington, New Zealand, 1/2-021592-G (retouched)**

that quite frequently not one fish was caught. Some of the fish in the early days were over 100 lb (45 kg), but by 1920 the average weight was 25 lb (11 kg) (Fig.102), and by the 1930s fish were generally less than 10 lb (4.5 kg).[156] Today, anecdotal evidence indicates that groper are no longer found in nearshore locations, and catch rates and average fish sizes have declined significantly, with large examples found only in shallow water at remote locations (Fig.103).[246]

Deregulation of the fishing industry in the 1960s and financial incentives brought in by the government to encourage investment led to rapid expansion of coastal fishing in inshore waters with large numbers of local vessels targeting popular species, while greatly increased foreign investment led to expansion of the deepwater fishing fleet. Over the next 15 years the industry grew dramatically, assisted by the development of modern, lightweight, expanded foam-insulated freezers, which could be carried on road vehicles, enabling distribution to a greatly expanded market and encouraging investment in the industry.

By the late 1970s some inshore fisheries were showing signs of overexploitation. For this reason no new rock lobster and scallop fishing permits were issued after 1978, and in 1980 a moratorium on issuing new licences for catching finfish was introduced. With too many boats chasing the same fish, catch rates declined and many fishers were forced out of business. Issues

FIG.103 **Today large groper are found in shallow coastal water only at remote locations such as Chatham Island.** © Matt Lind, Wildblue Spearfishing Ltd.

of biological and economic sustainability emerged as critical to the industry and industry restructuring was inevitable.[397]

Until the 1970s, fishing around New Zealand was primarily carried out by numerous single owner-operated vessels (Fig.107). These were generally small-scale enterprises restricted to inshore fisheries, with their catch being marketed to local communities. Overall, the fishing industry did not account for a large percentage of employment or revenue in the economy, with the exception of rural areas where fishing was frequently part-time, providing seasonal employment and income for many Māori when local freezing works were closed.[373] The combination of Britain's entry into the European Community with the associated loss of market access for New Zealand's agricultural products, and the OPEC oil shock of 1973, caused the New Zealand Government to introduce incentives aimed at expanding fisheries in an attempt to bolster the economy and increase exports.

Many commercial fisheries in New Zealand are tales of boom and bust, where uncontrolled exploitation resulted in a rapid decline of fish stocks, forcing fishers to target alternative, less desirable species, travel greater distances to fishing grounds, or venture further offshore to deeper waters to catch fish on an industrial scale. The decline in the fishery has often been slow and gradual, over many years, so that although the fishery was thought to be relatively robust, it could be severely depleted. These declines have often gone unnoticed because the overfished stocks are perceived as normal by each successive generation of fisheries scientists – a phenomenon known as shifting baselines – while the true decline is recognised only by customary, commercial and recreational fishers or researchers with decades of experience.[398-401]

Baseline estimates for many fish stocks in New Zealand today are derived from trawl surveys carried out only after the largely unregulated boom years of the 1960s and 1970s. Unregulated fishing during this period depleted the stocks,[146, 384] and many of New Zealand's fisheries were depressed to around, or even below, 20 percent of their unfished biomass.[384]

In the 1980s the Ministry of Fisheries recommended that the harvesting of rig or school shark (tupere, *Galeorhinus australis*) and elephantfish (reperepe, *Callorhinchus milii*) be reduced by up to 70 percent as catches progressively declined.[402, 403] Following analysis of historically reported fish catches to the Marine Department, and analysis of genetic diversity, fisheries researchers determined that stocks of snapper, one of the most popular commercial and recreational fish species taken in New Zealand waters, may have decreased by 80-95 percent in some regions around New Zealand between the 1930s and early 1990s,[145, 195] before beginning to partially recover after 1995 (see Fig.13).[401]

Despite this, some stocks failed to recover from historical overfishing, and some commercial catch quotas for snapper were cut in the late 1990s to protect the fish stock. Over the 1996 to 2009 period, the total allowable commercial catch for snapper declined by around 8 percent, although the reported catch since 2006 has remained stable.[404, 405] A stable level of catch, even over several years, has been shown to conceal the risk of sudden collapse of the fish stock.[406] In 2013 the Total Allowable Catch was increased with an additional 500 tonnes being allocated to recreational fishers (who took an estimated 45 percent of the TAC in 2011/12), but individual bag limits were reduced from ten to seven fish and the minimum size increased to 30 cm, while making no changes to the commercial catch.

Despite technological advances, especially in the locating of fish, and greater gear efficiency, the New Zealand fish catch peaked around 1998 and has since been declining steadily. In 2013 about 18 percent of all assessed stocks were deemed to be overfished (up from 15 percent in 2011) and required reductions in TAC to allow them to rebuild to target levels, while 6.5 percent of stocks were considered collapsed, requiring the complete closure of fisheries to rebuild the affected species. Almost 70 percent of all stocks were fished above the management target.[54]

The Quota Management System

In the 1980s New Zealand implemented the Quota Management System to control fishing and fish stocks. Based on Individual Transferable Quotas, the system was a radical departure from the previous fisheries management system and it quickly evolved to control the total commercial catch of all the main fish species found within New Zealand's 200-nautical mile Exclusive Economic Zone.

Concerns over the environmental effects of fishing increased in the last two decades of the 20th century, together with recognition that ongoing management must include consideration of the ecological impact on non-commercial as well as commercial fishes, as many species are of high importance to recreational and customary fishers.[384] Despite this, fisheries research funding

---

[54] Ministry for Primary Industries. (2013). The status of New Zealand's fisheries 2013. Retrieved from: http://www.mpi.govt.nz/document-vault/885

more than halved between the mid-1990s and early 2000s, while the list of commercially fished species brought into the Quota Management System increased from 26 to 96, essentially resulting in less and less being known about more and more.[407] Balancing protection and use of the marine resources required the government to properly fund and develop a more structured approach and involve all stakeholders in the process – tangata whenua, the fishing industry, environmental groups, recreational fishers, fisheries biologists, environmental scientists, economists, communities, and the wider public.

Under the Quota Management System, fish stocks are managed, usually by changing the allowable catch (TAC), so as to provide the maximum sustainable yield. The quantity of fish that can be taken from each fish stock by all fishers (commercial and non-commercial) is the Total Allowable Catch and each year an allocation is first made to provide for customary Māori use, then recreational fishing, and the remainder is made available to the commercial sector. This commercial catch can vary annually, and each commercial fisher's quota is a percentage of the total allowance for each species, not a fixed tonnage. The system operates on the assumption that the target set for each fish stock controls exploitation and prevents overfishing. It controls output (catch) rather than input (fishing effort), although poaching is a major issue for some stocks.[408]

The success of the Quota Management System can be attributed to New Zealand's geo-graphical isolation, with few fish stocks shared with other nations. The key fishing industry players, as well as fishery managers and politicians, were strong supporters and promoters of the system from its inception and were instrumental in getting the system introduced. The government also bought back catch histories from inshore fishers to reduce total allowable catches in stressed inshore stocks, so fishers were compensated for catch reductions, and at the time of catch reductions for inshore species, offshore, deepwater species were largely underexploited. Subsequently, catches from these latter fisheries increased, which formed a strong economic base for development of the New Zealand fishing industry.[409]

The general perception is that both the economic performance of the industry and biological status of the fishery resource have improved, with strategies in place to rebuild stocks.[384] The reality is that between 1996 and 2009 the economic performance of the industry increased because during this period the number of species in the Quota Management System increased from 26 to 96, but while the asset value for the original 26 species managed under the system increased by 18 percent, the total allowable commercial catch for these species was reduced by 41 percent,[405] suggesting an overall decline.

An open and transparent stock assessment and Total Allowable Catch setting process has evolved, with participation of user groups in fisheries management decision-making resulting in increased industry responsibility in the conduct of fishing operations. Bycatch problems, such as dumping of fish, appear to have been reduced by fishers adjusting their methods of operation and adopting industry codes of practice, and by the implementation of a system of overfishing provisions that has evolved to take into account changing circumstances.[409] It is acknowledged that many of New Zealand's fisheries resources were already depleted to around, and in some cases below, 20 percent of virgin (unfished) biomass prior to the introduction of the Quota Management System.[410] A number of fisheries in New Zealand still remain at very low stock levels by comparison with the virgin biomass, and the likelihood of these stocks ever being able to recover to unfished biomasses is often argued to be low.[384]

Current fisheries management philosophy considers overfishing to be a level of fishing that takes the resource below the state at which it is most productive, rather than the more general acceptance that there are fewer fish around than formerly. Management of many local fisheries is being undertaken without clear information on where the stock size is in relation to the level that would produce the maximum sustainable yield (MSY).[407, 411] This MSY concept has been painstakingly developed, intensely debated, and prematurely eulogised, and has ultimately evolved into a complex blend of language, mathematical theory and law.[412] Basic assumptions of maximum sustainable yield are that stocks can be managed outside their role in the ecosystem; that density dependence is the main regulating factor in population dynamics; and, that if there is enough information on the stock, then it is possible to fully control its trajectory.[413]

The flaw in this theory is that time-lags of decades to centuries can occur between the onset of overfishing and consequent changes in ecological communities, because unfished species of similar trophic level assume the ecological roles of overfished species until they too are overfished.[414, 415] Marine fisheries management has traditionally been based on the biology and population dynamics of individual target species, with management controls generally exercised through limits on individual fish sizes, seasons of harvest, catch limits, and restrictions on gear efficiency designed to protect reproductive stocks. For more than four decades, fatal flaws in single-species and population-based maximum sustainable yield approaches to fisheries management have been seriously discussed.[414, 416-422]

The premise behind the concept is that there are estimable levels of surplus production that may be safely removed from a given population (Fig.104). It has long been assumed that fish stocks and populations, and the ecosystems in which they exist, are healthy as long as they are maintained close to the levels or state that provide this yield. Nevertheless, a growing body of ecological, genetic and theoretical evidence suggests that this may not necessarily be so, for either the exploited species or their ecosystems. Errors in measuring stock yields lead to poor management decisions, and consequent changes to the life history of the target fish species can have unforeseen impacts. In addition, large-scale variations in productivity can occur naturally as a consequence of climatic variability affecting the natural balance within marine ecosystems. All these complicate management efforts to define the level of maximum sustainable yield (sometimes crudely interpreted as "harvest as much as possible"), often leaving fish stocks, and increasingly ecosystems, in jeopardy.[412] The MSY management strategy may have serious consequences for higher trophic level predators that rely on fish stocks for food, and could seriously impact populations of dolphins or seals with limited geographic ranges.

Broad-scale management by the Quota Management System does not necessarily recognise areas of local depletion or abundance. For example, localised areas of fish abundance in New Zealand waters, such as marine reserves, are excluded from the Ministry of Fisheries control, with management vested in the Department of Conservation. These reserves, or harvest refugia, should be evaluated as management tools to enhance or sustain coastal fisheries. Protected areas also safeguard the genetic diversity of wild stocks, serve as experimental controls for determinations of potential yield,[417] and can be used to determine natural and fishing mortality estimates.[423] Larvae and juvenile fish moving away from marine reserves make a significant contribution to adjacent fishing grounds. Studies on the Great Barrier Reef in Australia showed that a marine reserve covering 28 percent of a local reef area provided 50 percent of the recruited fish stock both within the reserve and up to 30 km distance.[424, 425] Whether this spillover fully compensates for the lost fishery or simply increases fishing effort in unprotected areas is unclear.[426-428]

Fisheries management in the 21st century

Approaches to fishery management continue to develop as understanding of the marine environment increases and attitudes change. There is now recognition that fisheries are part of an ecosystem and should not be managed in isolation, although concerns over the damage done to fisheries by practices such as bottom trawling have been expressed since 1376 AD.[429, 430] Government agencies and commercial industry groups are increasingly taking up concerns raised initially by environmental organisations.

In April 2007, the New Zealand Government decided to accept a proposal from the fishing industry to close 17 areas to bottom trawling and dredging, providing protection to an area of seabed habitat equal to 1.2 million square kilometres, approximately 30 percent of the EEZ – an area four times the landmass of New Zealand.[431, 432] Closing an area to bottom trawling is only a partial solution, as fish stocks on isolated seamounts can be targeted by longlining, or habitats can potentially be destroyed by offshore drilling and mining, and a trawling ban actually affords little overall protection. Conservationists have claimed that much of the closed area has only limited commercial fish stocks, or consists of rugged terrain, where trawling is impracticable if not impossible, and this has resulted in the accusation that the closure amounts to a cynical move by industry to look good while avoiding real concessions.[433]

With the exception of species with extremely limited distributions, such as Australian handfishes (Brachionichthyidae) which are found only in and near the Derwent estuary in

FIG.104 **Crew members preparing to empty a 25-tonne haul of orange roughy on board a factory trawler, 1981.**
© Peter McMillan, NIWA, Wellington.

Tasmania,[434] the only marine species mentioned in discussions of marine fish conservation prior to the 1990s have been large commercial species, threatened as a result of depleted stocks and overfishing. Species such as North Atlantic cod (*Gadus morhua*), bluefin tuna (*Thunnus thynnus*), or slow-breeding sharks may be in danger of extinction in some parts of the oceans,[435-438] and according to a report published in *Science Magazine* in 1998, the once common barndoor skate (*Raja laevis*), one of the largest skates in the northwest Atlantic, may have come close to extinction, or may have even become extinct at the end of the 20th century, without anyone having noticed.[439]

As a result of greater awareness of the effects of environmental degradation, alien species introductions and depletion of target species and bycatch stocks, an increasing number of coastal fish, particularly reef-dwelling species, are recognised as threatened or potentially threatened. Severe population declines have been documented for some grouper species (Serranidae) in the Atlantic and rockfishes (Sebastinae) in the Pacific, and for some sharks (Selachei), skates (Rajidae), and sawfishes (Pristidae).[436, 437] Several marine species are CITES[55] listed as threatened with extinction (CITES Appendix I), including coelacanth (*Latimeria* spp.), and an eastern Central Pacific species of sciaeanid or drum (*Totoaba macdonaldi*), or endangered (CITES Appendix II), including whale shark (*Rhincodon typus*), basking shark (*Cetorhinus maximus*), great white shark (*Carcharodon carcharias*), giant humphead wrasse (*Cheilinus undulatus*), and 42 seahorse species (*Hippocampus* spp.).

New Zealand fisheries are managed today by the Ministry of Primary Industries, which states: "We can boast perhaps the best managed fisheries in the world under the pioneering Quota Management System…" New Zealand is not immune from the global trends apparent, and wild fisheries are reaching their natural limits. The global fish catch (to which New Zealand contributes approximately 1 percent), which quadrupled between 1960 and 2000, is no longer rising, seemingly because oceanic fisheries cannot sustain a greater catch.[402, 415, 435]

The open-access approach to fisheries until the latter part of the 20th century led to predicted excess fishing effort, dissipation of economic profit from fishing, and inefficient depletion of fish populations.[440, 441] Throughout the world approximately 25 to 28 percent of the major fish stocks are thought to be in jeopardy of collapsing when estimates are based on scientific stock assessments;[442-444] however, when estimates are based on catch trends it has been suggested that around 70 percent of all stocks are overexploited.[427, 444, 445]

Fishery-related stock collapses appear to be increasing worldwide as the global removal of fish, invertebrates, discarded bycatch, illegal and unreported catch has reached an estimated 100-120 million tonnes annually over the last two decades.[406, 446] The problem is made worse as the

---

[55] Convention on International Trade in Endangered Species

aquatic eco-system is damaged by pollution and habitats destroyed, with some recent studies predicting a complete collapse of marine fisheries within 40 years.[447] There is mounting evidence that pelagic ecosystems could rapidly change from being fish-dominated to more rudimentary systems dominated by single-cell flagellate organisms and jellyfish, with lasting ecological, economic and social consequences.[446, 448]

Fishing industry proponents state that about 30 percent of the world's fish stocks are currently classified as overfished, but fishing pressure has been reduced enough so that all but 17 percent of stocks would be expected to recover to above overfished thresholds if current fishing pressure continues.[427] The idea that 70 percent of the world's fish stocks are overfished or collapsed or that the rate of overfishing is accelerating[445] was considered to be untrue by industry.[449, 450]

In 2005, the Ministry of Fisheries set out a new Strategy for Managing the Environmental Effects of Fishing (SMEEF). This document acknowledged that fishing had effects on the wider environment and described how limits must be set on the acceptable level of effect. The strategy included three key factors to be considered when setting environmental limits: long-term sustainability; a balance between use and protection; and the needs of future generations.[451]

All exploited coastal fish species have declined in abundance since colonial times. Too often, fisheries management is driven by market forces as a result of inadequate resourcing of management agencies, and at the expense of understanding the basic biological parameters of targeted species. Collapse of a fish stock usually results from complex scenarios, which include fluctuating environmental factors impacting ecosystems, impacts of land-based pollution and sedimentation, incorrect identification of fish species, inadequate biological and ecological knowledge of target species, and hasty management decisions such as overestimation of stock size, or proposed quotas that do not allow for natural declines in populations.

Fisheries issues include the target species as a bycatch in other fisheries, technological changes enabling the easier locating of fish, and greater gear efficiency. Socio-economic factors include quotas set above scientifically determined levels – often as a consequence of fishers demanding greater access to fish, politicians not responding to warnings from fisheries biologists and managers, illegal fishing, subsidies that enable fishers to continue fishing when it becomes uneconomic, as well as slow management response when initial signs of a collapse become apparent.[407, 452, 453]

Annual decisions on commercial catch limits by Ministry officials are often ambiguous, informal and vulnerable to outside influence from vested interests, resulting in some decisions on commercial fishing limits being described as "essentially guesswork".[407] Scientific objectivity within the management process is highly vulnerable because scientific decisions with significant policy implications are being made in a forum that is not designed to handle them.[454]

Research often targets only the most valuable species in the system, as they are of primary interest to fishing companies that may have contributed to the cost of the research. This results in resources being focused on a few species, with no incentive for research on more ecologically important issues or other minor species. Ignoring the lesser valued species is a risky option because the impacts on ecosystems could indirectly affect the more profitable species. A lack of understanding of the dynamics of marine ecosystems in New Zealand means that research agencies are unable to provide reliable advice, and without accurate scientific information the fisheries management structure risks being an elaborate house of cards.[454]

Over the last half of the 20th century, global fishing saw a gradual transition from targeting long-lived, high trophic level, piscivorous bottom fish toward short-lived, low trophic level invertebrates and planktivorous pelagic fish. Fishing down food webs (that is, at lower trophic levels) at first leads to increasing, then to stagnating or declining catches, indicating that present exploitation patterns are unsustainable.[455] Until the 1970s, fishing on the high seas was largely open access to any nation with the appropriate fishing vessel capacity, but today fishing is limited by 200-mile Economic Zones, and highly migratory fish stocks that move between different zones are overseen by 18 regional fisheries-management organisations. Decisions made by these consensus-oriented institutions, in which each member nation has equal status, are often guided more by politics and trading agreements than by sound science.

The purpose of the Fisheries Act is to provide for utilisation of fisheries resources while ensuring sustainability, and avoiding, reducing or mitigating adverse impacts of fishing on the marine environment. The Act does not define adverse effects; consequently, operational objectives cannot be defined and the trade-off between reaping of current benefits and avoiding long-term costs to the fishing industry threatens sustainability.[456]

Even sustainability itself is poorly understood in the marine environment. Criteria used to define endangered or threatened species in the terrestrial environment may differ significantly for marine species. A population decline of 50 percent in a population of a terrestrial species is regarded as a serious threat. In the marine environment a population decline of 50 percent is seen as a desirable management strategy to bring harvested stocks to a level that will result in maximum sustainable yield as the population attempts to recover.[416, 457]

Some marine species have been reduced much more than this. For example, in the 1960s the harvest of California pilchards (*Sardinops sagax caerulea*) declined to less than 5 percent of the catch taken in the 1930s, and it has been suggested that the population was reduced to less than 0.02 percent of the original stock size, but after almost half a century of protection the population appears to be recovering.[458, 459]

Concern by international environmental groups and retailers over depleted orange roughy stocks in the 1990s and early 2000s were dismissed as baseless by the New Zealand Ministry of Fisheries.[55] However, the Parliamentary Commissioner for the Environment expressed concerns for the orange roughy stock as early as 1992, stating: "…there is no doubt the fishery is at risk

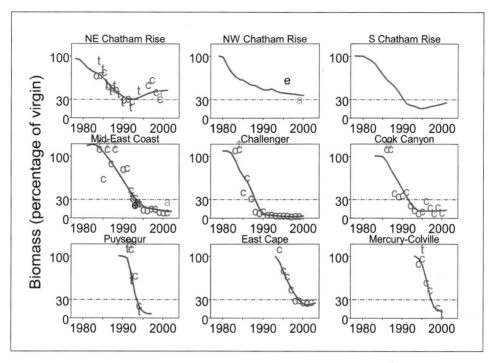

FIG.105 Orange roughy stock collapse in New Zealand waters. After Annala *et al.* 2003[521]

and that more courageous decisions are required if fisheries collapse is to be avoided…"[460]

In mid-2010 deepwater fisheries research scientists stated that assessments suggested the orange roughy fishery on the Chatham Rise was continuing to collapse, and some stocks on the Challenger Plateau and other areas were below 3 percent of original biomass, while other stocks that had been closed to fishing for 10 years had shown some limited signs of recovery (Fig.105).[461] Lack of good data on levels and patterns of recruitment in New Zealand waters is a major source of uncertainty in current stock assessments, and a principal concern for long-term sustainability of such fisheries.[462, 463]

There are few signs of biological compensation, and the recruitment levels of orange roughy appear to be low. In 2006, orange roughy became the first commercially fished species in Australasian waters to be added to the threatened species list by Australian authorities.[464] Despite this, the species is still taken commercially from Australian as well as New Zealand waters.

In 2007 New Zealand fisheries scientists expressed concern over declining stocks of hoki which had decreased to well below 20 percent of the virgin biomass as a result of low recruitment after 1995, and continued annual catches of 200,000 tonnes or more.[465] The hoki fishery was

---

Freshwater fisheries

New Zealand's freshwater fisheries are also under threat. One freshwater species, the grayling (upokorokoro, *Prototroctes oxyrhynchus*), once an important food resource for Māori,[1, 40] declined in abundance after the introduction of trout and other salmonids into New Zealand, and became extinct in the mid-20th century, for reasons that remain unclear.[174] Populations of native kokopu or whitebait (*Galaxias* spp.) have also been declining as a result of habitat loss.[174]

The longfin eel (*Anguilla dieffenbachii*) was an important food resource for pre-European Māori, as evidenced by the legends surrounding eel fishing (Fig.106), the construction of large eel weirs (Fig.14), and the hundreds of Māori names applied to ecological varieties and age classes of freshwater eels.[1, 175] Today the species is listed as 'At Risk (Declining)' by the Department of Conservation,[486] while at the same time it is managed as a commercial fish species by the Ministry of Primary Industries. Since the early 1990s the commercial harvest of eels has halved due to the rapidly declining population. Most longfin eels caught today are in the lowest size category (220-500g) and few large eels are seen.

Quota limits, used to control the catch and maintain longfin eel stock sizes, have been reduced each year as fishers fail to reach the set quota levels. Hence the stock is declining faster than the quota limits, and management has had no effect on preserving or conserving the stock, which is showing the classic signs of collapse – decline in average size and biomass,[487] a decline of recruitment by 75 percent between 1975 and 2004,[488] as well as reductions in geographic distribution, forcing fishers to find new places to fish, even into areas managed as conservation estate by the Department of Conservation.[443, 487, 489, 490]

Following an independent review of the fishery in 2013 by the Parliamentary Commissioner for the Environment,[57] the government announced a proposed package of measures designed to rebuild the population and improve long-term sustainability of longfin eel. This included a catch limits review, consideration of separating longfin and shortfin stocks in the South Island, and the introduction of abundance target levels, without supporting closure of the fishery.

certified as sustainable by the Marine Stewardship Council (MSC) in 2001.[466, 467] MSC is an independent, international, non-profit organisation set up to promote solutions to the problems being caused by unsustainable fishing practices. Between the 1997 and 2001 fishing years, the Total Allowable Commercial Catch (TACC) for hoki was set at 250,010 tonnes. In subsequent years, it was lowered progressively to 90,010 tonnes for the 2008 fishing year. These measures were aimed at rebuilding the fishery to a target level, and the TACC for the 2010 fishing year was

FIG.106 **Catching the legendary eel at Tangahoe. While eels of legendary size were uncommon, non-migratory female eels do occasionally attain lengths of up to 2m and were revered by Māori as gods (atua), feared as devils (tipua), or even fed and tamed, as well as being caught and eaten. Watercolour by Thomas William Downes, 1868-1938 (watercolour: Alexander Turnbull Library, Wellington, New Zealand, A-076-016)**

The Freshwater Fisheries Act gives complete protection to New Zealand's native fish with one exception: if fish are taken for 'human consumption' or for 'scientific purposes' then there is no protection. So in reality there is no practical, legal, or functional safeguard for our native freshwater fish. That fisheries and conservation managers permit destructive harvesting of species known to be critically endangered suggests adherence to a political ideology based on short-sighted economics, rather than sound ecological or long-term conservation principles.

[57] *On a pathway to extinction? An investigation into the status and management of the longfin eel* (April 2013; 94 pp.)

raised when stock assessments indicated that the fishery was within the target range. Subsequently, the TACC was raised to 120,010 tonnes for the 2011 fishing year.[468]

Lack of understanding of the underlying environmental processes and range of natural fluctuations under different environmental conditions makes it difficult to manage fisheries to ensure sustainability in a declining resource. Recent genetic studies have suggested that recruitment dynamics in marine species are more complex than previously assumed and indicate

FIG.107 Auckland waterfront, showing fishing boats alongside wharf, 1952. Whites Aviation Ltd :Photographs. Ref: WA-31160-F. Alexander Turnbull Library, Wellington, New Zealand. http://natlib.govt.nz/records/23261860

that estimated effective population sizes are two to six orders of magnitude smaller than census sizes,[469] which would have critical implications for fisheries management.

Ownership of quota is now concentrated in the hands of a few large companies, and their vessels and processing facilities are located at major ports (e.g. Bluff, Nelson, Dunedin) rather than in small coastal communities.[384, 474] While the majority of quota (by tonnage) is vested in these companies, a large proportion of fish is harvested by contract fishers, who purchase Annual Catch Entitlements (ACE) from quota owners (Fig.107). Hence the rights-holders are generally one step removed from the actual fishing operations.[384] The Fisheries Act grants rights to the individuals, companies or organisations holding quota. This enables the industry sector to have a higher influence over management decisions, and the uncertainty of science and research is not used as a reason for being cautious, but rather as a reason for *not* taking a more cautious approach. Government proposals to allow industry control of research funding and research objectives without independent monitoring can only exacerbate the situation.[456, 466]

Commercial fishing now extends further from New Zealand shores to encompass distant deepwater seamounts and, in the 1990s, even Antarctica, where signs of fishery collapse are already being hotly debated.[457, 475] The transferable quota management regime has been a major milestone on the pathway that ultimately aims to achieve sustainable domestic fisheries in New

Zealand;[384, 476] however, the system as presently practised will not necessarily ensure sustainable development of fisheries. After 20 years of the ITQ system in New Zealand, there is little evidence of rebuilding of many stocks. As indicated above, some stocks, including hoki, as well as several orange roughy stocks and snapper stocks, may be in an even worse state.[384, 456, 463, 466, 477]

Without adequate input from qualified fisheries researchers, accountability may be achievable only through public interest groups such as the Marine Stewardship Council certification of sustainable fisheries. Three New Zealand fisheries – hoki, albacore tuna (*Thunnus alalunga*), and Ross Sea toothfish (*Dissostichus eleginoides*) – have received MSC certification.[467, 478] Despite this, the sustainability of the hoki fishery has been questioned by fisheries managers[465] and toothfish stocks in McMurdo Sound and the southern Ross Sea have declined dramatically, according to some researchers.[457] Critics have suggested that the Marine Stewardship Council process does not fully take inadequate data into consideration and that several fisheries, including Antarctic toothfish, have been certified incorrectly.[479, 480]

New Zealand is ranked among the 10 best nations for management of their fishing industries in terms of progress in implementing ecosystem-based management of fisheries,[481] but in assessments of management effectiveness, or fisheries management performance in terms of biodiversity and aquaculture, no country is rated overall as 'good' and only four countries are regarded as 'adequate'.[482, 483] Reports suggest that in most instances worldwide where a total allowable catch has been the prime or sole method of fisheries management, the result has not been effective stock conservation, or even optimal economic utilisation of the resource.[473]

Reports also indicate that in most instances worldwide where a total allowable catch has been the prime or sole method of management, stocks and harvests have declined significantly, and some stocks have done so under the system in New Zealand,[145, 470-473] with at least 15 percent of the country's fish stocks estimated to be overexploited in the early 2000s.[484] Recreational fishers have pointed out that the lack of funding and research in some provincial areas has resulted in fish stocks being in a poor state – in Hawke's Bay, 94 percent of recreational, customary, commercial and charter fishers surveyed in 2012 considered that the availability of finfish had decreased over the last five years.[58]

Management decisions on commercial fishing limits in New Zealand have been described by fisheries science researchers as essentially guesswork and highly susceptible to influence, while political pressure from wealthy quota-holders has undermined the transparency and accountability of decision-making and science, and has also created resistance against broader ecosystem issues.[384, 407, 485] While there has been improvement on many fronts, the future is unclear, especially regarding the sustainability of the fish.

---

[58] Wayne Bicknell & Rich Burch, *State of Hawkes Bay Fisheries*, Guardians of Hawke Bay Fishery, Press Release, 11 December 2012

10

Mana tangata: *fishing rights*

"The Natives occasionally exercise certain privileges or rights over tidal lands. They are not considered as the common property of all Natives in the colony; but certain hapus or tribes have the right to fish over one mudflat and other Natives over another. Sometimes even this goes so far as to give certain rights out to sea. For instance, at Katikati harbour, one tribe of Natives have the right to fish within the line of tide-rip; another tribe of Natives have the right to fish outside the tide-rip. The lands contained in the schedule of the bill are probably the most famous patiki (flat fish) ground in New Zealand, and have been the subject of fighting between various hapus of the Thames Natives..."

James Mackay, 1862 (Mackay 1862: 7)

The Māori belief system was area-based and traditionally involved a complex arrangement of nested rights and responsibilities involving extended families (whānau), villages (kāinga), tribes (iwi), and subtribes (hapū), specifying who could fish and when, where, and how, enforced by formal and informal cultural norms, beliefs, institutions, and ritual.[485]

Early settlers recorded numerous anecdotal comments on the abundance of fish around the New Zealand coast, and how Māori harvested them. They described how several tribes would frequently gather together to harvest fish using huge nets or fleets of canoes, and the thousands of fish taken that were eaten fresh or sun-dried on racks for winter supplies. However, many of the European settlers equated fishing with poverty, a subsistence pastime for Māori.

Māori culture changed as new crops and livestock became available, and although their fishing activity declined significantly between 1800 and 1900, it continues today as part of Māori life (Fig.109, 110). The history of 140 years of commercial fisheries in New Zealand from the signing of the Treaty of Waitangi in 1840 until the introduction of the Quota Management System in the 1980s is largely a tale of open access to what was seen as an unlimited resource.

New Zealand's coastal harbours were rapidly developed by European settlers and traders in the 19th century because of the ease of access the waterways provided to large areas of hinterland in all weather (Fig.108). Following the sealing and whaling days, the first Europeans ventured inland in the early 1800s, seeking new economic opportunities such as timber, particularly kahikatea and kauri (*Agathis australis*) for their sailing ships, and other resources such as gold. The sawmills flourished and large quantities of timber – both sawn and logs – were exported by ship. Land-based whaling stations were established in the 1820s in harbours close to the whale migration routes through Foveaux Strait and Cook Strait.[88] By 1840 European settlers had begun to establish farms and a number of settlements developed around suitable harbours.[491]

Toward the end of the 19th century accessible timber resources were close to being exhausted and the goldfields were in decline. Emphasis in trade shifted to the export of kauri gum in the north, and wool and frozen meat from pastureland throughout the colony. The Europeans began to look for other industries to develop, and fishing became increasingly important. As European interest in commercial fishing developed, Māori fishing property rights were acknowledged as an issue, but were regarded as separate from commercial development.

From the earliest contact with Europeans, Māori had shown an eagerness to trade and in

FIG.108 **Akaroa Harbour, 1840. Lithograph of a painting by Louis Auguste Marie Le Breton, 1818-1866. Alexander Turnbull Library, Wellington, New Zealand, PUBL-0028-185**

FIG.109 Māori communal fishing at Parengarenga, North Auckland, 1907. Christchurch City Libraries, Christchurch, Photo CD8, IMG0053

doing so, they dominated commerce prior to 1840. They achieved considerable commercial success with exports to Australia – in 1830, 28 ships averaging 110 tons made 56 voyages between New Zealand and Sydney carrying Māori-grown potatoes and milled grain as well as timber, flax, pork, and dried fish.[305] The flourishing trade enabled Māori to obtain European goods, including muskets.[492] Tribes with the advantage of muskets as weapons (occasionally assisted by Europeans), attacked their traditional enemies. The resulting 'Musket Wars' from around 1810 to the early 1830s decimated many tribes, with an estimated 20,000 deaths between 1820 and 1830 alone.[493-495]

In October 1831, Captain Laplace, of the French exploratory vessel *La Favorite*, arrived at the Bay of Islands to rest and refresh his crew. Relationships between the French crew and local Māori were not good, and rumours spread that the French vessel, with 400 men on board, had come for the purpose of seizing the country and avenging the deaths of Marion du Fresne and his fellow crew members from the vessel *Mascarin*, who had been massacred nearby in 1772. In response a number of northern chiefs (encouraged by the British missionaries), petitioned the English King William IV (1765-1837) for an economic and defence alliance to protect them from the French, and to control the lawlessness of the English sealers and sailors. In 1833 James Busby was

FIG.110 **Three unidentified Māori girls shelling toheroa on a beach, ca 1910-39. Northwood brothers :Photographs of Northland. Alexander Turnbull Library, Wellington, New Zealand, Ref: 1/1-026522-G**

appointed by the Governor of New South Wales in Australia as the first official representative of the British Government in New Zealand.[305] Busby reported that a state of anarchy existed in the country as an ever-increasing number of British settlers were arriving.

Although some settlers and British politicians wanted to protect and safeguard Māori rights, others wanted to open the country for wider settlement.[496] In a British parliamentary debate in the House of Commons on the state of New Zealand (which was seen as 'The Southern England of Future Ages'), the Hon. Mr Hawes (Member for Lambeth), described how Busby proposed to bring about a confederation of the chiefs and to govern this confederation himself, in a manner that "…the power of the Natives would be nominal only…". Hawes described this scheme of asserting and upholding British authority under the pretence of maintaining a nominal Native Power as impracticable, and "…scarcely an honest one…"[496] The leading missionaries in New Zealand, who themselves had become speculators in land, were also alarmed by a possible French dominion after rumours that a settler, Baron Charles de Thierry, was going to set up an independent state at Hokianga to bring in the French, and they joined with Busby and other settlers in the Bay of Islands in desiring the establishment of a national power in the country.[82]

Busby convened a meeting of 35 northern Māori chiefs in 1834 to discuss difficulties facing

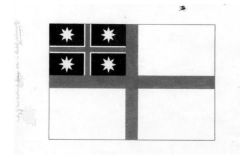

FIG.111 **The flag of the United Tribes of New Zealand,
the nation's first flag, together with a declaration of an
independent state, was accepted and acknowledged
by the British Crown (King William IV, 1830-37) in 1835.
Later, the Treaty of Waitangi was signed between the
British Crown and 'the chiefs of the United Tribes of
New Zealand' in recognition of their independent
sovereignty, which continued after 1840. United Tribes
Ensign copied from a plate in the Admiralty Library's
Book of flags, 1845. This reproduction by James
Laurenson has a white fimbriation in place of the black
fimbriation of the original. Alexander Turnbull Library,
Wellington, New Zealand, MS-Papers-0009-09-01**

their trading vessels in international waters and Sydney Customs regulations after a Hokianga-built ship had been seized in Sydney for not flying a national flag. The following year a declaration of independence was signed by the Confederation of United Chiefs who formed themselves into an independent state, with the title of 'The United Tribes of New Zealand'. They invited the southern tribes to join the Confederation as one nation and agreed to meet in Congress for the purpose of framing laws for the dispensation of justice, the preservation of peace and good order, and the regulation of trade. They also chose the first national flag (Fig.111).[305] Edward Jerningham Wakefield, Secretary of The New Zealand Company which had been formed as a commercial operation to obtain land from Māori (often under dubious circumstances) and to assist immigrants from Britain, noted that:

"…there cannot be the least doubt that this document was composed by the missionaries at the Bay of Islands, and signed by the chiefs with as little real comprehension of its meaning as had attended the signature by natives of the deeds of feoffment (a grant of lands as a fee) drawn up by Sydney attorneys with blanks for the names of places…"[305]

Māori maintained their autonomy in good faith and continued to seek economic opportunities. The declaration, printed and published in 1836 and 1837, was acknowledged by King William and ratified by the British Government through the House of Commons; however, it was never taken seriously until it proved to be an impediment to the annexation of New Zealand.[497]

The Treaty of Waitangi

In 1839 William Hobson was sent by the British Government to negotiate with Māori to revoke the declaration of independence, and pave the way for annexation of New Zealand, allowing the transmission of sovereignty to Queen Victoria. The Treaty of Waitangi 1840, negotiated by Hobson, guaranteed Māori "…full exclusive and undisturbed possession of their… Fisheries…" (Fig.112)[498]

FIG.112 Māori and Pākehā had different expectations following the signing of the Treaty of Waitangi, 1840. *Evening Post* cartoon by Tom Scott, 1965. Alexander Turnbull Library, Wellington, New Zealand, J-065-038

Although Māori understood the arrangement to be a partnership with reciprocal obligation, and mutual benefit with the guarantee of tino rangatiratanga (governance or sovereignty) ensuring autonomy over their own affairs,[305] the influx of British settlers expanded exponentially and a new settler government was established with the Constitution Act in 1852.[499] The new colonial governing structure actively repressed Māori political aspirations, denied Māori access to educational resources, political power and commercial opportunities, offered no influential allies, and violently repressed Māori resistance to British settlement.[420]

Māori rapidly developed a range of shared political objectives to halt settlement. The settlers' antagonistic attitude toward Māori and seemingly insatiable demand for land culminated in the New Zealand Land Wars of the 1860s. In contrast to the Land Wars, for 30 years following the signing of the Treaty of Waitangi, fisheries remained relatively undeveloped by Europeans, and Māori fishing rights were not disputed.

Commercial fisheries development

The commercial exploitation of New Zealand fisheries by the Europeans was opened up in

1877 when Chief Justice Prendergast ruled that the Treaty of Waitangi was a "legal nullity". The Treaty had guaranteed Māori the right to keep their fisheries; hence, transformation of Māori rights from the position of aboriginal title subsisting at law, to one held on sufferance of the Crown,[500] was essential before fisheries could be made available for commercial exploitation by the settlers. In Prendergast's view, New Zealand was peopled only by "…primitive barbarians and savages…" who had no sovereignty to cede nor existing body of customary law that could be legally recognised,[501] and therefore those Māori proprietary rights – including fisheries – confirmed in the Treaty were unenforceable against the Crown in the New Zealand courts. This view reflected the views of many of the colonists at the time and was based on popular belief of the ultimate sovereignty of parliament and held that the Treaty would be legally binding only if enshrined in legislation.[420, 502, 503]

From the first legislative interventions in the form of the Oyster Fisheries Act 1866, and the Fish Protection Act 1877 – passed without any consultation with Māori – it became clear that the Crown assumed it had the right to restrict or deny access to commercial fisheries by

FIG.113 **The *Dominion*, operated as a gillnet boat by the Famularo family, photographed at Island Bay in the 1930s by Sydney Charles Smith. Alexander Turnbull Library, Wellington, New Zealand, ID 1/4-021121-G (retouched)**

both Māori and non-Māori alike. In response to a request from the Hon. Captain Fraser in the House of Representatives, the Speaker, the Hon. William Fitzherbert, ruled that the 1877 Bill did not need to be printed in Māori as it did not "specifically" affect them.[504] Both domestic and commercial harvesting of oysters became illegal without a licence, and all commercial fisheries were subject to controls on the type of nets and other fishing equipment that could be used.[91]

The government of the day viewed Māori involvement in the fisheries, along with oyster-gathering, as a politically sensitive issue, and showed a paternalistic view toward Māori. Secretary for Marine, W.T. Glasgow, thought it "…very unfair to allow the Māoris [sic] to deplete the fisheries… in their own interests they should be protected from the effects of their ignorance and improvidence…", although at the same time he was wary of any interference with their "…supposed rights…"[65] Despite this, the government seemed to have no practical grasp of how new laws would affect traditional Māori fishing.[374, 505]

The New Zealand Government legislative intervention in sea fisheries was a result of concerns for the conservation and exploitation of resources, and the legislation provided for the general public exploitation of fisheries resources, based on the premise of the Crown's right to provide for this, notwithstanding the rights guaranteed to Māori under the Treaty of Waitangi.[91, 93] As a consequence, no effort was made to consult with Māori before exercising legislative control over fisheries. In 1877 the Fish Protection Act established the Marine Department to administer new regulations for all waters within three miles of the coast. This was followed by the Fisheries Conservation Act 1884, which required Māori to obtain a licence when fishing for any reason other than for personal or family consumption.[506]

The Fisheries Encouragement Bill 1885 provided for the establishment of fishing villages and promoted the production of smoked and canned fish for export. The Commissioner of Trade and Customs, R.A.A. Sherrin, described the intent of the Bill as "…free fishing for fishermen and their descendants, the tenure that of the impersonal crown, free from the caprice of individuals; the sea swarms with fish open to men of all nations…" (Fig.113).[88]

Māori throughout New Zealand protested strongly. In the last three decades of the 19th century and into first decade of the 20th, some 46 Māori fishing petitions were referred to the Native Affairs Committee, with the complaints also directed to the loss of control by the chiefs, of the mana of the lands, forests and fisheries, which had been guaranteed by the Treaty.[91, 93] In 1870, the Māori Land Court observed:

"…the use to which the Maoris [sic] appropriated this land [coastal foreshore] was to them to the highest value no one acquainted with their customs and manner of living can doubt. It is very apparent that a place which afforded at all times, and with little labour and preparation, a large and constant supply of almost the only animal food which they could obtain, was of the greatest possible value to them; indeed of very much greater value and importance to their existence than any equal portion of land on terra firma…"[502]

Europeans blamed Māori for the decline of fish stocks, particularly the mullet resource in Northland. In a Marine Department report to the New Zealand House of Representatives (Parliament) dated 1895, it was stated:

"…Representations having been made to the department that it would be desirable to prescribe a close season for mullet in all waters between Cape Wiwiki and the North Cape, and also to prohibit the Maoris [sic] from using certain methods of fishing which had the effects of depleting the fishery, in consequence of their taking small mullet in large quantities it is recommended that… Maoris [sic] be made amenable to the fishery regulations…"[507]

On 21 December 1896 regulations under the Sea Fisheries Act 1894 set the style of nets and minimum mesh sizes according to non-Māori standards.[91] Māori fishing activity was diminished by requiring a larger mesh than they had traditionally used, and their involvement in the commercial fishery was further reduced as non-Māori fishing increased.

Since fisheries did not appear to be as profitable as agriculture and farming, the government generally did not pay much attention to fishing.[373] The Oyster Fisheries Act of 1892, and the Sea Fisheries Acts of 1894 and 1903 extended control over commercial fisheries and introduced new restrictions, appointed Fisheries Officers and Inspectors, as well as introducing new licensing, reporting and vessel registration requirements. Thus, although the government supported fishery exploitation and condoned free market operations, there was no significant effort to promote or encourage commercial fisheries other than through limited subsidies on exports of smoked or canned fish.[146, 338]

The Whitebait Fisheries Regulations 1894 and 1896 continued the change in the fishing rights and management responsibilities of Māori. When sea fisheries were first made the subject of statutory regulation in the Fish Protection Act 1877, Māori rights under the Treaty of Waitangi were preserved. This provision was omitted in 1894 and reinstated in 1903 in a vaguer form. From 1900 to 1962 Māori fishing rights were protected under law which allowed fishing grounds to be reserved on application to the Marine Department in particular areas for meeting personal needs. Although the statutory provision was in force for 62 years, no applications were ever approved.[91]

These Acts were introduced under the false assumption that the dramatically increased pressure on aquatic resources by European settlers could be managed separately and have little

FIG.114 Unidentified Māori woman gathering seafood (probably oysters or mussels) from the rocks, possibly in Paihia. Photograph taken by Frank Denton between 1895 and 1900. Alexander Turnbull Library, Wellington, New Zealand, 1/2-008468-G

or no effect on the customary fisheries of Māori, and created an unnatural division not only between commercial and customary fisheries, but also between the English settlers and indigenous Māori.[485]

Although containing various clauses about Māori rights not being affected by the Acts, the successive pieces of fisheries legislation limited customary Māori fisheries and banned commercial exploitation of many indigenous fish species on Māori terms, limiting them to fishing for personal needs only (Fig.114). The Whitebait Fisheries Regulations 1894 and 1896 specifically outlawed the Māori trapping techniques of building weirs and diversion channels. Almost all the court cases in which Māori continued to assert their customary rights or Treaty rights against the Crown were dealt with as criminal prosecutions, with Māori fishers defending themselves against convictions under the Fisheries Acts for fishing in a customary manner outside Pākehā law.[508] These cases became more common in the early 20th century as Pākehā and Māori society came closer together and Crown control over fisheries was extended.

The final nail in the coffin of Māori attempts to have their fishing rights recognised in the courts came in 1914 when a Māori woman convicted for whitebaiting with an illegal net

appealed her conviction in the Supreme Court on the grounds that she was exercising a Māori fishing right as protected under fisheries legislation.[509] In the absence of any legislation defining such a right, her case was rejected and the Court ruled that Māori therefore were bound by the provisions of the Fisheries Act 1908. This judgement effectively eliminated all customary Māori fishing rights until the Treaty of Waitangi Act of 1975 recognised the Treaty in law. The Waitangi Tribunal was set up in 1975 to investigate Māori grievances, but initially could only make recommendations (which were not legally binding) on events dating from after the establishment of the Tribunal.[503]

Declaration of an Exclusive Economic Zone

In 1977 New Zealand joined the increasing number of countries claiming sovereign rights over their coastal waters. On the basis of international agreements signed at the Third United Nations Conference on the Law of the Sea, the following year New Zealand established a 200-mile Exclusive Economic Zone (the world's fourth largest at 1.2 million square nautical miles).[373] The government contributed to overuse of this resource by subsidising entry into the fishing industry and encouraging fishery exploitation through duty-free imports, loans, investment allowances, export tax incentives and price supports, as well as encouraging joint venture arrangements with foreign companies (Fig.116). In addition to rapidly increasing fishing technology and expertise, government policies also provided wider access to international markets. The policy changes countered a long-standing Māori culture of taking only what was needed and having a diversified livelihood beyond fishing. Thus, it became hard for part-time fishers to compete with an increasingly large and technologically sophisticated fishing industry.[373]

The Fisheries Act 1983 was passed in order to consolidate and update regulations which had been in force since 1908[93] and to help address issues associated with the general decline of inshore fish stocks in the 1960s and 1970s following deregulation of the industry in 1963, and the exploitation of deepwater stocks following the declaration of the 200-mile Exclusive Economic Zone in 1978 which had seen high levels of foreign fishing throughout the preceding decades (Fig.115). Although the 1983 Act acknowledged Māori fishing rights, it did little to implement them, and deliberately excluded many part-time fishermen from the industry, many of whom were Māori, and halved the number of permit holders.[146]

Following the common property trend of overexploitation, commercial fleets continued to invest heavily, resulting in widespread overcapitalisation in the market. The areas of greatest overfishing were along the eastern and northern coasts of the North Island, precisely where large populations of Māori lived. In 1983, the government's efforts to curtail overfishing of inshore fisheries involved the involuntary exclusion of fishers earning less than 80 percent of their income or $10,000 per year from fishing (estimated at around 8000 individuals).[373, 420, 510] Beyond doing little to reduce overcapitalisation, this ill-conceived policy had a disproportionately adverse effect on Māori.

FIG.116 *Akebono Maru*, the largest factory trawler to have fished New Zealand waters, moored in Timaru Harbour, 1984. © Peter McMillan, NIWA, Wellington.

In 1985 the Waitangi Tribunal's jurisdiction and mandate was widened to cover claims (including the loss of fisheries) prior to 1975 and extending back to 1840, and the settlement of grievances became a major focus for Māori.[503]

In 1986, the 1983 Fisheries Act was amended to introduce the Quota Management System (QMS), which provided a new way of managing commercial fish stocks.[511] Prior to this, clear evidence of Māori traditional practices and Treaty rights was presented to Waitangi Tribunal hearings between 1982 and 1984 on Waitara and Manukau fisheries.[91, 512] Despite these hearings, which vindicated Māori claims, the government proceeded with its unilateral imposition of transferable rights to commercial fisheries in defiance of the Treaty claims.[513]

Under the QMS, rights to harvest set quantities of certain commercial fish species were allocated to fishermen who had commercial fishing permits, based on their historical catch record (Fig.117). As many Māori and other part-time fishers had been excluded from the industry in 1983 and no longer had commercial fishing permits, they did not receive Quota allocations. Māori groups challenged the QMS through the Courts and the Waitangi Tribunal, claiming that the QMS was unfair because it ignored Māori fishing rights guaranteed under the Treaty of Waitangi.[91-93]

In 1988, the New Zealand Attorney-General Geoffrey Palmer declared that the Treaty was part of the Law of the Land. In both High Court and Court of Appeal judgements, Treaty standing was renewed.[420] The court system bolstered the Tribunal's emphasis on the Treaty in the *Te Weehi v. Regional Fisheries Officer* decision of 1986. Fisheries officers arrested Te Weehi, a North Island Māori man living in the South Island, for taking undersized pāua. He appealed his conviction in the High Court, arguing that he had asked South Island elders for permission to take the shellfish and his Treaty rights overrode conservation regulations. The court ruled in his

FIG.117 **Fishing boat entering Wellington Heads. Fishing Quotas were issued to commercial fishers based on historical catch records, but ownership of Quota is now held by a few large companies and most fish are harvested by contract fishers who purchase Annual Catch Entitlements from Quota owners.** © Chris Paulin.

favour with similar reasoning, thus reintroducing the Treaty into domestic law after a 140-year absence, and forcing the government to acknowledge Treaty claims.[420, 513]

In upholding the challenge and ordering the government to negotiate with Māori, the High Court opened the way for a relatively small group of tribal representatives to gain a significant stake in the New Zealand fishing industry on behalf of their people. The *National Business Review* envisaged proposals for a Māori-owned company that would account for almost 39 percent of the NZ$1.5 billion annual earnings of the fishing industry.[514] The High Court decision and subsequent negotiations also set the scene for a significant economic and political empowerment of these tribal leaders at the expense of urban, non-tribal Māori leaders and their constituencies. At each successive stage in the acrimonious litigious and political struggle that ensued, the paramount legitimacy of indigenous tribes (iwi) and their representatives was affirmed and strengthened.

The New Zealand Government eventually acknowledged Māori customary and commercial fishing as an integral part of fisheries management. The Māori Fisheries Act 1989 was introduced as an interim solution, recognising customary rights, buying back 10 percent of Quota already allocated and reserving 20 percent of all future allocation for Māori.[485, 511] The historic Deed of Settlement, signed in late 1992, provided an agreement in which the Crown funded Māori in a 50/50 joint venture with Brierley Investments Ltd to bid for Sealord Products Ltd – New Zealand's largest fishing company. In return, Māori agreed that all current and future claims in respect of commercial fishing rights were fully satisfied and discharged.[420, 511, 515]

A five-year review process, begun in 1991, resulted in the Fisheries Act 1996, which introduced regulations that defined how customary fishing could take place and the rights and responsibilities of tangata whenua in managing their own customary Māori fisheries. Later, the Kaimoana Customary Fishing Regulations 1998 and the Fisheries (South Island Customary Fishing) Regulations 1998 strengthened some of the rights of tangata whenua to manage their fisheries. These regulations let iwi and hapū manage their non-commercial fishing in a way that best fits their local practices, without having a major effect on the fishing rights of others.[509] When the government sets the total catch limits for fisheries each year, it allows for this customary use of fisheries.

A governance body, the Treaty of Waitangi Fisheries Commission Te Ohu Kai Moana, was set up in 2004 to oversee all Māori commercial fishing settlement assets including developing the resource (Te Wai Māori), and educating Māori (Te Putea Whakatupu) in the industry.[476, 511] Since 1992 the value of these assets has tripled, and now represents around a third of New Zealand's commercial fishing industry, administered under a company called Aotearoa Fisheries.

To use the customary fishing regulations, iwi and hapū groups must decide who has tangata whenua status over a fishery (this can be shared by a number of groups). The tangata whenua then nominate people to act as guardians for the area (tangata kaitiaki in the North and Chatham Islands, tangata tiaki in the South and Stewart Islands) who are then appointed by the Minister of Fisheries. Guardians can issue permits, allowing people to catch fish in their area for customary use. There is no cap on customary harvest; however, catches are reported to the Ministry of Fisheries so that regulators take account of the customary use when catch limits are set each year. An allowance is also made for the estimated recreational fishing harvest, which is also uncapped but unreported (Fig.118).

Tangata whenua can also request special management areas – mātaitai reserves and taiāpure local fisheries – to manage some of their traditional fishing grounds. Within mātaitai reserves, guardians set rules for customary and recreational fishing, and can say whether some types of commercial fishing should continue within the reserve.[509] While the legislation allows management of the fish stock, it does not provide protection of the habitat the fishery relies on.[516]

Other fishing method restrictions and closures for customary purposes are also available under section 186A of the Fisheries Act. Without the customary fishing regulations, iwi and hapū can take fish only for important events through Regulation 27 of the Amateur Fishing Regulations. This lets marae honour guests by providing seafood at events like hui and blessings, but it gives no other control over their fisheries than this.[517]

The Quota Management System has its supporters and critics. Although it has been instrumental in managing commercial fish stocks, it has been described as a corporate model that may be causing fish to be 'mined', with the high initial proceeds invested in other ventures and in lobbying to protect short-term interests.[485] Any proportional reduction in annual quota share for management purposes does not give rise to compensation for fishers.[385]

Ko Aotearoa Tēnei (2011), a report produced by the Waitangi Tribunal into claims concerning New Zealand law and policy affecting Māori culture and identity (referred to as Wai 262), recommended reforms to relevant laws and policies so that the interests of kaitiaki (guardianship) can be fairly and transparently considered alongside other interests, as neither the Crown nor Māori have ownership rights over New Zealand's indigenous species, but there are obligations for both parties under the Treaty to protect resources surrounding those species.[147]

The Crown asserts that it has sovereignty or authority to own and manage resources in the interests of conservation and sustainability under the concept of kāwanatanga conferred by Article 1 of the Treaty of Waitangi. Māori assert that their tino rangatiratanga or full authority,

FIG.118 The *Auckland Weekly News* captioned this 1901 photograph 'New Zealand sea fishing: A fine basket of hāpuka. Fifteen excursionists fishing off Gannet Island caught in a few hours 40 hāpuka, weighing an average of 50 lb each, the largest scaling 90 lb'. Sir George Grey Special Collections, Auckland Libraries, AWNS-19010118-5-2

guaranteed by Article 2, affords them ownership and kaitiaki management rights over these same resources.

Māori claims to commercial fisheries were settled with the Treaty of Waitangi (Fisheries Claims) Settlement Act 1992. This Act resolved commercial fishing issues and ended the Waitangi Tribunal's involvement in Māori fishing claims and the Tribunal Fisheries Commission was disbanded. The Māori Fisheries Act 2004 provided for the implementation of the 1992 Deed of Settlement and the allocation and management of aquaculture settlement assets. An agreement in principle was reached in 2008 which covered all major fish and shellfish species farmed at that time. Today Māori have advanced fisheries scholarships and training, processing and marketing interests, and there are now Māori representatives on fishery statutory bodies and Māori-appointed guardians of exclusive fishing areas or mātaitai.[485]

By negotiating with only a few influential Māori and leaving the question of allocation to the Treaty of Waitangi Fisheries Commission, composed of those influential Māori involved with the negotiations, the government undermined Māori collective activity regarding fisheries and left them to confront the Commission, and to argue amongst themselves, and through the High Court over material benefits, including the role of traditional tribes and whether urban Māori should be considered tribes for the purposes of distributing assets.[420, 518, 519] In the process, authority (tino rangatiratanga) of individual Māori groups was usurped by the Commission and Māori lost access to the Waitangi Tribunal and the courts for all future fishing concerns.[420, 520]

Adjustments to the legislative scheme have been ongoing. Today the Quota Management System does not grant full ownership of the fish themselves as property, as the rights granted are in fishing, and ownership remains with the Crown. The Crown retains the right of protection, management, and most importantly, the right to create and correspondingly to extinguish fish quota by legislation in the interests of conservation and sustainability, thus managing fisheries resources and meeting the concept of kāwanatanga in Article 1 of the Treaty.[385]

The fisheries Treaty settlement resulted in enormous benefits for Māori, but Māori tribes, individuals, and fishermen have not yet attained an equitable settlement which fully meets kaitiaki obligations of tino rangatiratanga under Article 2 of the Treaty. Under the Treaty principles the Crown has a duty to actively protect Māori interests when introducing new legislation.

The Exclusive Economic Zone and Continental Shelf (Environmental Effects) Act 2012 set up a legislative framework for environmental management in New Zealand's Exclusive Economic Zone and Continental Shelf, and introduced a requirement for the Crown to consult with Māori. This will enable Māori to have a stronger voice in governance, resource management, and resource planning, while at the same time allowing for customary fishing activities. The opportunity provided by investment in collaboration and co-governance should ensure a sustainable future for New Zealand's marine environment.

Āhua momo tuhituhi: *literature cited*

1. Best, E., (1929a). *Fishing methods and devices of the Maori. Dominion Museum Bulletin no. 12*, Wellington: W.A.G. Skinner, Govt. Printer. viii, 230 pp.

2. Paulin, C.D., (2007). Perspectives of Māori fishing history and techniques: Nga āhua me nga pūrākau me nga hangarau ika o te Māori. *Tuhinga – Records of the Museum of New Zealand Te Papa Tongarewa* 18: 11-47.

3. Beaglehole, J.C., (1942). *Abel Janszoon Tasman & the discovery of New Zealand*, Wellington: Department of Internal Affairs. 66 pp.

4. Royal, T.A.C., (1992). *Te Haurapu: An introduction to researching tribal histories and traditions*, Wellington: Bridget Williams Books. 111 pp.

5. Barber, I., (2003). Sea, land and fish: spatial relationships and the archaeology of South Island Maori fishing. *World Archaeology* 35(3): 434-448.

6. Best, E., (1929b). *The Whare Kohanga (the "nest house") and its lore. Dominion Museum Bulletin*, Wellington, N.Z.: Government Printer. 72 pp.

7. Cowan, J., (1930). *The Maori: Yesterday and Today*, Christchurch: Whitcombe and Tombs. 266 pp.

8. Poata, T.H., (1919). *The Maori as a Fisherman and his methods*, Opotiki: W.B. Scott & Sons. 26 pp.

9. Anderson, A., (1983). *When all the moa ovens grew cold*, Dunedin: Otago Heritage Books 52 pp.

10. Anderson, A., (1997). Uniformity and regional variation in marine fish catches from prehistoric New Zealand. *Asian Perspectives: the Journal of Archaeology for Asia and the Pacific* 36: 1-26.

11. Anderson, A. and J.H. Beattie, (1994). *Traditional lifeways of the Southern Maori. The Otago University Museum ethnological project*, Dunedin: Otago University Press. 640 pp.

12. Nagaoka, L.A., (2002). The effects of resource depression on foraging efficiency, diet breadth, and patch use in southern New Zealand. *Journal of Anthropological Archaeology* 21: 419-442.

13. Beattie, H., (1939). *Tikao talks: Ka taoka o te ao kohatu. Treasures from the ancient world of the Maori*, Auckland: Penguin Books. 160 pp.

14. Tikao, T.T., (1939). *Tikao talks: Traditions and tales told by Teone Taare Tikao to Herries Beattie*, Wellington: A.H. & A.W. Reed. 23-50 pp.

15. Best, E., (1976a). *Maori religion and mythology: being an account of the cosmogony, anthropogeny, religious beliefs and rites, magic and folk lore of the Maori folk of New Zealand*, Wellington: Govt. Printer. 2 vols., 424, 684 pp.

16. Chapman, G.T., (1892). On the working of Greenstone or Nephrite by the Maoris. *Transactions and Proceedings of the New Zealand Institute* 24: 479-539.

17. Beaglehole, J.C., (1955). *The voyage of the Endeavour 1768-1771*, Cambridge: The Hakluyt Society. 696 pp.

18. Beaglehole, J.C., (1961). *The journals of Captain Cook. II. The Voyage of the Resolution and Adventure 1772-1775*, Cambridge: University Press. 1021 pp.

19. Beaglehole, J.C., (1962). *The Endeavour journal of Joseph Banks 1768-1771*, Sydney: Angus and Robertson. 114 pp.

20. Beaglehole, J.C., (1967). *The journals of Captain Cook. III. The Voyage of the Resolution and Discovery 1776-1780*, Cambridge: University Press. 2 vols., 720, 928 pp.

21. Forster, G., (1777). *A voyage around the World in His Majesty's Sloop, Resolution, Commanded by Captain James Cook, during the Years 1772, 3, 4, and 5*, London: B. White, J. Robson, P. Elmsly, and G. Robinson. 2 vols., 602, 607 pp.

22. Parkinson, S., (1784). *A journal of a voyage to the South Seas, in his Majesty's ship the Endeavour*, London: Dilly & Phillips. 212 pp.

23. Dieffenbach, E., (1843). *Travels in New Zealand*, London: John Murray. 396 pp.

24. Colenso, W., (1869). On the Maori races of New Zealand. *Transactions and Proceedings of the New Zealand Institute* 1: 1-75.

25. Colenso, W., (1891). Vestiges: Reminiscences Memorabilia of works, deeds, and sayings of the ancient Maoris. *Transactions and Proceedings of the New Zealand Institute* 24: 445-467.

26. Buck, P.H., (1929). *The coming of the Maori*. 2nd ed, Nelson, N.Z.: Cawthron Institute. 43 pp.

27. Worthy, T.H. and R.N. Holdaway, (2002). *The Lost World of the Moa*, Bloomington: Indiana University Press. 718 pp.

28. Smith, I.W.G., (1989). *Maori impact on the marine megafauna: pre-European distributions of New Zealand sea mammals*, in *Saying so doesn't make it so: papers in honour of B. Foss Leach*, D.G. Sutton, Editor, New Zealand Archaeological Association Monograph. p. 76-108.

29. Holdaway, R.N. and C.J. Jacomb, (2000). Rapid extinction of the moas (Aves: Dinorinthiformes): Model, test, and implications. *Science* 287: 2250-2254.

30. Lalas, C. and C.J.A. Bradshaw, (2001). Folklore and chimerical numbers: review of a millennium of interaction between fur seals and humans in the New Zealand region. *New Zealand Journal of Marine and Freshwater Research* 35: 477-497.

31. Leach, B.F., (1989). Archaeological time trends in South Island Maori fishing. *Museum of New Zealand Te Papa Tongarewa Technical Report* 11: 1-38.

32. Petchey, F. and T. Higham, (2000). Bone diagenesis and radiocarbon dating of fish bones at the Shag River mouth site, New Zealand. *Journal of Archaeological Science* 27(2): 135-150.

33. McGlone, M., Anderson, A.J. and Holdaway, R.N., ed. (1994). *An ecological approach to the Polynesian settlement of New Zealand.* The origins of the first New Zealanders, ed. D.G. Sutton. Auckland University Press: Auckland. 136-163, 270 pp.

34. Daniel, M.J., (1990). Order Chiroptera, in *The handbook of New Zealand mammals.* p. 114-137 in C.M. King, Editor. Auckland: Oxford University Press.

35. Best, E., (1924b). *The Maori as he was: a brief account of Maori life as it was in pre-European days*, Wellington, N.Z.: Government Printer. xv, 295 pp.

36. Davidson, J.M., (1984). *The Prehistory of New Zealand*, Auckland: Longman Paul. 270 pp.

37. Hiroa, T.R., (1949). *The coming of the Maori.*, Wellington: Maori Purposes Fund Board. 551 pp.

38. Firth, R., (1929). *Economics of the New Zealand Maori*, Wellington: R.E. Owen 519 pp.

39. Travers, W.T.L., (1873). On the life and times of Te Rauparaha. *Transactions and Proceedings of the New Zealand Institute* 5: 1-222.

40. Best, E., (1903). Food products of Tuhoeland: being notes on the food-supplies of a non-agricultural tribe of the natives of New Zealand; together with some account of various customs, superstitions, &c., pertaining to foods. *Transactions and Proceedings of the New Zealand Institute* 35: 44-111.

41. Buck, P.H., (1921). Maori food-supplies of Lake Rotorua, with methods of obtaining them, and usages and customs appertaining thereto. *Transactions and Proceedings of the New Zealand Institute* 21: 433-451.

42. Salmond, A., (2003). *The Trial of the Cannibal Dog, Captain Cook in the South Seas*, London: Penguin Press. 506 pp.

43. Nagaoka, L.A., (2001). Using diversity indices to measure changes in prey choice at the Shag River mouth site, southern New Zealand. *International Journal of Osteoarchaeology* 11: 101-111.

44. Leach, B.F., (2006). Fishing in Pre-European new [sic] Zealand. *New Zealand Journal of Archaeology Special Publication and Archaeofauna* 15: 1-359.

45. Davidson, J., *et al.*, (2002). Pre-European Māori Fishing at Foxton, Manawatu, New Zealand. *New Zealand Journal of Archaeology* 22: 75-90.

46. Leach, B.F. and A.S. Boocock, (1979a). Prehistoric man in Palliser Bay. *National Museum of New Zealand Bulletin* 21: 272.

47. Rick, T.C. and J. Erlandson, (2008). *Human Impacts on Ancient Marine Ecosystems: A Global Perspective*, Berkeley and Los Angeles: University of California Press. 336 pp.

48. Cassels, R., ed. (1984). *The role of prehistoric man in the faunal extinctions of New Zealand and other Pacific Islands.* Quaternary Extinctions, ed. P.S.A.K. Martin, R. G. University of Arizona: Tucson 741-767. In: 892 pp.

49. Wilmshurst, J.M., *et al.*, (2004). Early Maori settlement impacts in northern coastal Taranaki, New Zealand. *New Zealand Journal of Ecology* 28(2): 167-179.

50. Rawlence, N.J., *et al.*, (2012). Soft-tissue specimens from pre-European extinct birds of New Zealand. *Journal of the Royal Society of New Zealand* iFirst: 1-28.

51. Buck, P.H., (1926a). *The Maori craft of netting. Transactions of the New Zealand Institute*, Wellington: Govt. Print. 597-646 pp.

52. Thomson, J.T., (1867). *Rambles with a philosopher; or views at the Antipodes by an Otagonian*, Dunedin: Mills, Dick & Co. 250 pp.

53. Best, E., (1924a). *The Maori. Memoirs of the Polynesian Society*, Wellington, N.Z.: Board of Maori Ethnological Research for the Author and on behalf of the Polynesian Society. 2 vols., 530, 628 pp.

54. Leach, B.F. and A.S. Boocock, (1993). Prehistoric fish catches in New Zealand. *British Archaeological Reports International Series* 584: 1-38.

55. Leach, B.F., Quinn, C., Morrison, J. and Lyon, G., (2003). The use of multiple isotope signatures in reconstructing prehistoric human diet from archaeological bone from the Pacific and New Zealand. *New Zealand Journal of Archaeology* 23: 31-98.

56. Poata, T.H., (1921). The Art of Fishing. Some Acient Customs and Traditions. *Evening Post* (21 December 1921): 13.

57. Best, E., (1919a). *The land of Tara and they who settled it: the story of the occupation of Te Whanga-nui-a-Tara (the great harbour of Tara), or Port Nicholson, by the Maoris*, New Plymouth, N.Z.: Printed for the [Polynesian] Society by Thomas Avery. 121 pp.

58. Firth, R., (1926). Proverbs in Native Life, with Special Reference to Those of the Maori, II. (Continued). *Folklore* 37(3): 245-270.

59. Nicholas, J.L., (1817). *Narrative of a voyage to New Zealand performed in the years 1814-1815, in company with the Reverend Samuel Marsden, Principal Chaplain of New South Wales.* Vol. 1, London: James Black and Sons. 303 pp.

60. Savage, J., (1807). *Some account of New Zealand*, London: Union Printing Office. 110 pp.

61. Anderson, A., (1981). Barracouta fishing in prehistoric and early historic New Zealand. *Journal de la Socieété des Océanie* 37: 145-158.

62. Salmond, A., (1991). *Two worlds: First meetings between Maori and Europeans* 1642-1772, Auckland: Viking. 477 pp.

63. Hoare, M.E., ed. (1982). *The Resolution Journal of Johann Reinhold Forster 1772-1775. Volume 2.* Vol. 2. The Hakluyt Society: London. 182 pp.

64. Cruise, R.A., (1824). *Journal of a ten months residence in New Zealand*, London: Longman, Hurst, Rees, Orme, Browne and Green. 321 pp.

65. Johnson, D., (2004). *Hooked, the story of the New Zealand fishing industry*, Christchurch: Hazard Press. 551 pp.

66. Ollivier, I. and C. Hingley, (1982). *Early eyewitness accounts of Maori life 1. Extracts from journals relating to the visit to New Zealand of the French ship St Jean Baptiste, in December 1769 under the command of J.F.M. de Surville*, Wellington: Alexander Turnbull Library Endowment Trust in Association with the National Library of New Zealand. 225 pp.

67. Polack, J.S., (1838). *New Zealand: being a narrative of travels and adventures during a residence in that country between the years 1831 and 1837*, London: Richard Bentley. 149 pp.

68. Magra, J., (1771). *A Journal of a Voyage round the World in his Majesty's ship Endeavour*, London: T. Beckett and P.A. de Hondt. 275 pp.

69. Crozet, J.M., (1891). *Crozet's Voyage to Tasmania, New Zealand, the Ladrone Islands, and the Philippines in the years 1771-1772, translated by H. Ling Roth.*, London: Truslove & Shirley. 148 pp.

70. Affairs, D.o.M., (1965). The first Pakehas to visit the Bay of Islands.

Te Ao Hau (June 1965): 14-18.

71. Dunmore, J., (1965). *French explorers in the Pacific*, Oxford: Clarendon Press. 428 pp.

72. Smith, S.P., (1909). Captain Dumont d'Urville's visit to Whangarei, Waitemata, and the Thames in 1827. *Transactions and Proceedings of the Royal Society of New Zealand* 42: 412-433.

73. Wright, O., (1955). *The voyage of the Astrolabe – 1840: an English rendering of the journals of Dumont d'Urville and his officers of their visit to New Zealand in 1840, together with some account of Bishop Pompallier and Charles, Baron de Thierry*, Wellington: Reed. 180 pp.

74. Hector, J., (1884). The fisheries of New Zealand. *Bulletin of the United States Fish Commission* 4: 53-55.

75. Brunner, T., (1959). *Journal of an expedition to explore the interior of Middle Island, New Zealand, 1846-8. Early travellers in New Zealand*, ed. N.M. Taylor, Oxford: Clarendon Press. 192-211, in 594 pp.

76. Yate, W., (1835). *An Account of New Zealand; and of the formation and progress of the Church Missionary Society's mission in the northern island*, London: Seeley & Burnside. 164 pp.

77. Buller, R.J., (1878). *Forty years in New Zealand: including a personal narrative, an account of Maoridom, and of the christianization and colonization of the country*, London: Hodder and Stoughton. 503 pp.

78. Elder, J.R., (1932). *Letters and Journals of Samuel Marsden*, Dunedin: Coulls, Somerville and Wilkie Ltd. 280 pp.

79. Earp, G.B., (1853). *New Zealand: its emigration and goldfields*, London George Routledge and Co. 260 pp.

80. Tregear, E., (1885). *The Aryan Maori*, Wellington: G. Didsbury, Government Printer. 107 pp.

81. Tregear, E., (1904). *The Maori Race*, Wanganui: A.D. Willis. 592 pp.

82. Wakefield, E.J., (1845). *Adventure in New Zealand From 1839 to 1844 With Some Account of the Beginning of the British Colonization of the Islands*, John Murray, Albemarle Street. 568 pp.

83. Ward, J., (1840). *Supplementary information relative to New Zealand, compiled for the use of Colonists*, London: New Zealand Company. 168 pp.

84. Heaphy, C., (1880). Notes on Port Nicholson and the natives in 1839. *Transactions and Proceedings of the New Zealand Institute* 12: 33-39.

85. Mair, W.G., (1873). Notes on Rurima Rocks. *Transactions and Proceedings of the New Zealand Institute* 5: 151-153.

86. Matthews, R.H., (1911). Reminiscences of Maori life fifty years ago. *Transactions and Proceedings of the New Zealand Institute* 43: 598-604.

87. Downes, T.W., (1918). Notes on eels and eel weirs (Tuna and Patuna). *Transactions and Proceedings of the New Zealand Institute* 50: 296-316.

88. Sherrin, R.A.A., (1886). *Handbook of the fishes of New Zealand*, Auckland: Wilson and Horton. 307 pp.

89. Taylor, R., (1855). *Te Ika a Maui, or New Zealand and its inhabitants, illustrating the origin, manners, customs, mythology, religion, rites, songs, proverbs, fables, and language of the natives. Together with the geology, natural history, productions, and climate of the country; its state as regards Christianity; sketches of the principal chiefs, and their present position*, London: Wertheim & Macintosh. 490 pp.

90. Maori Affairs Department, (1958). From a notebook by an unknown Arawa of last century. *Te Au Hou The New World* 22(6): 50.

91. Rata, M. and Others, (1988). *Wai 22 – Report of the Waitangi Tribunal on the Muriwhenua fishing claim*, Wellington Waitangi Tribunal, New Zealand Department of Justice. 371 pp.

92. Waitangi Tribunal (1988). *Report of the Waitangi Tribunal on the Muriwhenua fishing claim, Wai 22. Wai*, Wellington, N.Z.: The Tribunal. xxi, 370 pp.

93. Waitangi Tribunal (1992). *The Ngāi Tahu sea fisheries report, 1992. (Wai 27)* Wellington: Brooker and Friend. 400 pp.

94. Ward, A., (1974). *A show of justice: racial "Amalgamation" in nineteenth century New Zealand*, Auckland: Auckland University Press. 382 pp.

95. O'Regan, T., (1992). Old myths and new politics: some contemporary uses of traditional history. *New Zealand Journal of History* 26(1): 5-27.

96. Baucke, J.F.W., (1905). *Where the white man treads: selected from a series of articles contributed to 'The New Zealand Herald' and 'The Auckland Weekly News'; including others published for the first time* Auckland: Wilson and Horton. 307 pp.

97. Grace, A.A., (1901). *Tales of a dying race*, London: Chatto & Windus. 250 pp.

98. Hill, H., (1902). The Maoris to-day and to-morrow. *Transactions and Proceedings of the New Zealand Institute* 35: 169-186.

99. Manning, F.E., (1863). *Old New Zealand, a tale of the good old days by a Pakeha maori, of Anthropophagi, and men whose heads grow between their shoulders*, Auckland: Robert J, Creighton & Alfred Scales. 350 pp.

100. Simmons, A., (1879). *Old England and New Zealand: the government, laws, churches, public institutions and the resources of New Zealand, popularly and critically compared with those of the Old Country: with an historical sketch of the Maori race (the natives of New Zealand): to which are added extracts from the author's diary of his voyage to New Zealand, in company with 500 emigrants*, London: Stanford. 143 pp.

101. Hamilton, A., (1908). Fishing and sea foods of the Maori. *Dominion Museum Bulletin* 2: 1-73.

102. Hiroa, T.R., (Buck, P.H.) (1921). Maori food supplies of Lake Rotorua, with methods of obtaining them, and usages and customs appertaining thereto. *Transactions and Proceedings of the New Zealand Institute* 53: 432-451.

103. Hiroa, T.R., (Buck, P.H.) (1926). The Maori craft of netting. *Transactions and Proceedings of the New Zealand Institute* 56: 597-646.

104. Marshall, Y., (1987). Maori mass capture of freshwater eels: an ethnoarchaeological reconstruction of prehistoric subsistence and social behaviour. *New Zealand Journal of Archaeology* 9: 55-79.

105. Travers, W.T.L., (1872). Notes upon the historical value of the 'Traditions of the New Zealanders', as collected by Sir George Grey, K.C.B., the late governor-in-chief of New Zealand. *Transactions and Proceedings of the New Zealand Institute* 4: 51-62.

106. Wohlers, J.F.H., (1875). The mythology and traditions of the Maori in New Zealand. *Transactions and Proceedings of the New Zealand Institute* 7: 690-692.

107. Best, E., (1919b). *A Maori fish-trap showing unusually fine workmanship*, Wellington, N.Z.: Govt. Printer. 35-37 pp.

108. Best, E., (1976b). *The Maori canoe: an account of various types of vessels used by the Maori of New Zealand in former times, with some description of those of the Isles of the Pacific, and a brief account of the peopling of New Zealand. Dominion Museum bulletin no. 7*, Wellington: A.R. Shearer, Govt. printer. 452 p. pp.

109. Beattie, H., (1920). Nature-lore of the southern Maori *Transactions and Proceedings of the New Zealand Institute* 52: 53-77.

110. Grey, G., (1928). *Nga mahi a nga tupuna*, New Plymouth Thomas Avery & Sons. 211 pp.

111. Peart, J.D., (1937). *Old Tasman Bay: a story of the early Maori of the Nelson District, and its association with Europeans prior to 1842, supplemented with a list of native place names*, Nelson R. Lucas. 143 pp.

112. Durie, M.H., (1998). *Te mana, te kawanatanga: the politics of Māori self-determination*, Auckland: Oxford University Press. 280 pp.

113. Johansen, J.P., (1958). Studies in Maori rites and myths. *Historisk-filosofiske meddelelser* 37(4): 1-201.

114. Ritchie, J., (1992). *Becoming Bicultural*, Wellington Huia Publishers Daphne Brasell Associates Press. 214 pp.

115. Stack, J.W., (1878). Sketch of the traditional history of the South Island Maoris. *Transactions and Proceedings of the New Zealand Institute* 10: 57-92.

116. Smith, S.P., (1910). *History and traditions of the Maoris of the West Coast, North Island of New Zealand, prior to 1840*, New Plymouth: Polynesian Society. 562 pp.

117. Lysnar, F.B., (1915). *New Zealand: the dear old Maori land*, Auckland Brett Printing and Publishing. 268 pp.

118. Dickison, M., (2009). The asymmetry between science and traditional knowledge. *Journal of the Royal Society of New Zealand* 39(4): 171-172.

119. Widdowson, F. and A. Howard. *Aboriginal "Traditional Knowledge", Science and Public policy: Ten Years of listening to the Silence.* (2006). Available from: http://blogs.mtroyal.ca/fwiddowson/files/2009/10/Aboriginal-traditional-knowledge-science-and-public-policy-Widdowson-and-Howard-peer-review-copy1.pdf.

120. Grey, G., (1854). *Ko nga moteatea, me nga hakirara o nga Maori*, Wellington: The Honorable Robert Stokes. 202 pp.

121. Smith, S.P., (1900). The Tohunga-Maori: a sketch. *Transactions and Proceedings of the New Zealand Institute* 32: 253-270.

122. McRae, J., (2000). *Māori Oral Tradition Meets the Book*, in *A book in the hand: essays on the history of the book in New Zealand.* p. 1-14 in P. Griffith, P. Hughes, and A. Loney, Editors. Wellington: Auckland University Press, 272 pp.

123. White, J., (1887-1891). *The ancient history of the Maori, his mythology and traditions*. Vol. 1-6, Wellington: George Didsbury, Government printer pp.

124. Ryan, P.M., (1995). *The Reed Dictionary of Modern Māori*, Auckland: Reed. 648 pp.

125. Johannes, R.E., (1981). *Words of the lagoon: fishing and marine lore in the Palau district of Micronesia*, Los Angeles: University of California Press. 245 pp.

126. Leach, B.F., (2003). *Depletion and loss of the customary fishery of Ngāti Hinewaka: 130 years of struggle to protect a resource guaranteed under Article Two of the Treaty of Waitangi. Report presented to the Waitangi Tribunal.* Vol. Document Wai 863-#A. 71 pp.

127. Macdonald, C., Penfold, M. and Williams, B. (eds) (1991). *The book of New Zealand women: ko kui ma te kaupapa*, Wellington: Bridget Williams Books. 772 pp.

128. Beattie, J.H., ed. (1994). *Traditional Lifeways of the Southern Maori: The Otago University Museum Ethnological Project, 1920.* ed. J. Anderson. University of Otago Press in association with Otago Museum: Dunedin. 636 pp.

129. Dacker, B., (1990). *The people of the place: mahika kai*, Wellington: New Zealand 1990 Commission. 39 pp.

130. Williams, J., (2009). "O ye of little faith": Traditional knowledge and Western science. *Journal of the Royal Society of New Zealand* 39: 167-169.

131. Chambers, C., (2009). Mixing methodologies: the politics of research techniques. *Journal of the Royal Society of New Zealand* 39(197-199).

132. Moller, H., *et al.*, (2009). Traditional ecological knowledge and scientific inference of prey availability: harvests of sooty shearwater (*Puffinus griseus*) chicks by Rakiura Māori. *New Zealand Journal of Zoology* 36: 259-274 36: 259-274.

133. Binney, J., (1987). Māori oral narratives, Pākeha written texts: two forms of telling history. *New Zealand Journal of History* 21: 16-28.

134. Dalley, B. and J. Phillips, (2001). *Going public: the changing face of New Zealand history*, Auckland: Auckland University Press. 226 pp.

135. Barber, E.W. and P.T. Barber, (2004). *When they Severed Earth from Sky: How the Human Mind Shapes Myth*, Princeton: Princeton University Press. 312 pp.

136. Cashman, K.V. and S.J. Cronin, (2008). Welcoming a monster to the world: Myths, oral tradition, and modern societal response to volcanic disasters. *Journal of Volcanology and Geothermal Research* 176: 407-418.

137. McWethy, D.B., *et al.*, (2009). Rapid deforestation of South Island, New Zealand, by early Polynesian fires. *The Holocene* 19: 883-897.

138. Wilmshurst, J.M., *et al.*, (2008). Dating the late prehistoric dispersal of Polynesians to New Zealand using the commensal Pacific rat. *Proceedings of the National Academy of Sciences of the United States of America* 105: 7676-7680.

139. Leach, B.F. *Archaeozoology in New Zealand*. (1998). Available from: http://www.cs.otago.ac.nz/research/foss/Archaeozoology/archzoo.htm.

140. Beasley, H.G., (1928). *Pacific Island Records: Fishhooks*, London Seeley, Service and Co. 133 pp.

141. Watt, R.J., (1990). *The fake Māori artefacts of James Edward Little and James Frank Robieson*. Unpublished PhD Thesis, Wellington: Victoria University 2 vols., 350, 370 pp.

142. Paulin, C.D., (2010). Māori fishhooks in European Museums. *Tuhinga Records of the Museum of New Zealand* 21: 13-41.

143. Mair, W.G., (1923). *Reminiscences and Maori stories*, Auckland: The Brett Printing & Publishing Co. 120 pp.

144. Craig, E.W.G., (1964). *Man of the Mist – a Biography of Elsdon Best*, Wellington: A.H. & A.W. Reed. 247 pp.

145. Annala, J.H., (comp.) (1994). *Report from the Fishery Assessment Plenary, May 1994: stock assessments and yield estimates. Unpublished report held in MAF Fisheries Greta Point library*, Wellington. 242 pp.

146. Paulin, C.D. and L.J. Paul, (2006). The Kaipara mullet fishery: nineteenth-century management issues revisited. *Tuhinga Records of the Museum of New Zealand Te Papa Tongarewa* 17: 1-26.

147. Tribunal, W., (2011). *Ko Aotearoa Tēnei: a report into claims concerning New Zealand law and policy affecting Māori culture and identity. Te Taumata tuatahi*. Wai 262, Wellington: Waitangi Tribunal. 257 pp.

148. Wallace, P.T.A., (2008). Exploring the Interface of Science and Mātauranga Māori. *Transformations '07: Composing the nation: ideas, peoples, histories, languages, cultures, economies*, Victoria University, Wellington.

149. Park, J.W., ed. (2005). Surimi seafood: products, market, and manufacturing. *Surimi and Surimi Seafood*, ed. J.W. Park. CRC Press Boca Raton. Pp. 375-433 in 885 pp.

150. Anderson, A.A.J. and D.G. Sutton, (1973). Archaeology of Mapoutahi Pa, Otago *New Zealand Archaeological Association Newsletter* 16(3): 107-118.

151. Leach, H.M. and G.E. Hamel, (1978). The Place of Taiaroa Head and Other Classic Maori Sites in the Prehistory of East Otago. *Journal of the Royal Society of New Zealand* 8(3): 239-251.

152. Trotter, M.M., (1967). Excavations at Katiki Point, *Records of the Canterbury Museum* 8(3): 231-245.

153. Leach, B.F. and A. Anderson, (1979b). The role of labrid fish in prehistoric economics in New Zealand. *Journal of Archaeological Science* 6(1): 1-15.

154. Higham, T.F.G. and P.L. Horn, (2000). Seasonal dating using fish otoliths: Results from the Shag River mouth site, New Zealand *Journal of Archaeological Science* 27(5): 439-448.

155. Paul, L.J., (1986). *New Zealand fishes: an identification guide*, Auckland, N.Z.: Reed Methuen. viii, 184 pp.

156. Graham, D.H. and G.P. Whitley, (1956). *A Treasury of New Zealand Fishes*. 2nd ed, Wellington, N.Z.: Reed. 424 pp.

157. Mehl, J.H.P., (1971). Spawning and length-weight of barracouta (Teleosti: Gempylidae) from eastern Cook Strait. *New Zealand Journal of Marine and Freshwater Research* 5: 300-317.

158. Travers, W.T.L., (1869). On the Changes Effected in the Natural Features of a New Country by the Introduction of Civilized Races. *Transactions and Proceedings of the New Zealand Institute* 2: 299-330.

159. Paulin, R., (1889). *The Wild West Coast of New Zealand: a summer cruise in the Rosa*, London: Thorburn & Co. 121 pp.

160. Hutton, F.W., (1904). *Index faunae Novae Zealandiae*, London: Dulau & Co. 398 pp.

161. Thomson, A.S., (1859). *The story of New Zealand: past and present, savage and civilized*, London John Murray; Christchurch Capper Press. 2 vols., 311, 368 pp.

162. Hector, J., (1874). Anniversary Address of the President. *Transactions and Proceedings of the New Zealand Institute* 6: 367-376.

163. Paulin, C.D. and J.H. Atkinson, (1984). *A key to families of New Zealand bony fishes. Miscellaneous series/National Museum of New Zealand*, Wellington, N.Z.: National Museum of New Zealand. 30 pp.

164. Paulin, C.D., *et al.*, (1989). *New Zealand fish: a complete guide. National Museum of New Zealand miscellaneous series*, Wellington, N.Z.: GP Books. xiv, 279 pp.

165. Roberts, C.D., Paulin, C.D., Stewart A.L. McPhee R.P. and R.M. McDowall ed. (2009). Checklist of living lancelets, jawless fishes, cartilaginous fishes and bony fishes. *The New Zealand Inventory of Biodiversity. Volume 1. Kingdom Animalia*, ed. D.P. Gordon. Vol. 1. Canterbury University Press: Christchurch. 529-538 in: 568 pp.

166. Stewart, A. and M. Clarke, eds. (2010). Beyond light – the great unknown. *Deep: Talks and Thoughts Celebrating Diversity in New Zealand's Untouched Kermadecs*, ed. A. Connell. Pew Environment Group and Te Papa Tongarewa: Wellington. Pp 69-70 in 92 pp.

167. Best, E., (1924c). Maori religion and mythology: being an account of the cosmogony, anthropogeny, religious beliefs and rites, magic and folk lore of the Maori folk of New Zealand: Section I. *Dominion Museum Bulletin.*, Wellington: Govt. Printer. 264 pp.

168. McDowall, R., (2010). New Zealand's distinctive and well known freshwater fish fauna. *Fish and Fisheries* 32: 1-33.

169. Pond, W., (1997). The land with all woods and water. p. 79-105 in H. Riseborough and J. Hutton, Editors: *Waitangi Tribunal Rangahaua Whanui Series*.

170. McDowall, R., (2011). *Ikawai: freshwater fishes in Māori culture and economy* Christchurch: University of Canterbury Press. 832 pp.

171. McDowall, R., (1984). *The Whitebait Book*, Wellington: A.W. & A.H. Reed. 210 pp.

172. O'Donnell, E., (1929). *Te Hekenga: early days in Horowhenua, Being the Reminiscences of Mr. Rod McDonald*, New Plymouth: G.H. Bennett & Co. Ltd. 138 pp.

173. Martin, W., (1929). *The fauna of New Zealand*, Christchurch: Whitcombe & Tombs. 266 pp.

174. McDowall, R.M., (1990). *New Zealand freshwater fishes: a natural history and guide*, Auckland: Hienemann Reed. 553 pp.

175. Strickland, R.R., (1990). Nga tini a Tangaroa: a Māori-English, English-Māori dictionary of fish names. *New Zealand fisheries occasional publication* 5: 1-64.

176. McDowall, R., (1991). Conservation and management of the whitebait fishery. *New Zealand Department of Conservation Science and Research Series* 38: 18.

177. Whitley, G.P., (1968). A checklist of the fishes recorded from the New Zealand region. *Australian Zoologist* 15(2): 1-102.

178. Ayling, T. and G. Cox, (1982). *Collins Guide to the Sea Fishes of New Zealand*, Auckland: Collins. 343 pp.

179. Doogue, R.B. and J.M. Moreland, (1966). *New Zealand Sea Anglers' Guide*, Wellington: Reed. 317 pp.

180. Ngata, H.M., (1993). *English-Māori Dictionary*, Wellington: Learning Media. 257 pp.

181. Williams, H.W., (1971). *A Dictionary of the Māori Language*, Wellington: Government Printer. 253 pp.

182. Paulin, C.D., (1982). Scorpionfishes of New Zealand (Pisces; Scorpaenidae). *New Zealand Journal of Zoology* 9: 437-450.

183. Pickard, C. and C. Bonsall, (2004). Deep-Sea Fishing in the European Mesolithic: Fact or Fantasy? *European Journal of Archaeology* 7(3): 273-290.

184. De Leo, F.C., *et al.*, (2010). Submarine canyons: hotspots of benthic biomass and productivity in the deep sea, *Proceedings of the Royal Society B: Biological Sciences* 277(1695): 2783-2792.

185. Golson, J., ed. (1959). *Culture change in prehistoric New Zealand*. Anthropology in the South Seas, ed. J.D. Freeman and W.R. Geddes. Avery Press: New Plymouth. 267 pp.

186. Leach, F. and A.S. Boocock, (1993). *Prehistoric fish catches in New Zealand. BAR international series*, Oxford: Tempus Reparatum. 38 pp.

187. Nordhoff, C., (1930). Notes on the off-shore fishing of the Society Islands. *Journal of the Polynesian Society* 39: 1-79.

188. Gudger, E.W., (1927). *Wooden hooks used for catching sharks and Ruvettus in the South Seas: A study of their variation and distribution*, New York: Museum of Natural History. 150 pp.

189. Kennedy, D.G., (1929). Field Notes on the Culture of Vaitupu, Ellice Islands. *Journal of the Polynesian Society* 38: 1-99.

190. Macgregor, G., (1937). Ethnology of Tokelau Islands. *Bernice P. Bishop Museum Bulletin* 146: 1-183.

191. Rainbird, P., ed. (2007). The role of fishing lure shanks for the past people of Pohnpei, eastern Caroline Islands, Micronesia. *Vastly ingenious: the archaeology of Pacific material culture*, ed. A. Anderson, K. Green, and F. Leach. Otago University Press: Dunedin. 217-226, 319 pp.

192. Nakamura, I., ed. (1995). Gempylidae. *Escolares. Guia FAO para Identificación de Especies para lo Fines de la Pesca. Pacífico Centro-Oriental*, ed. F.K. W. Fischer, W. Schneider, C. Sommer, K.E. Carpenter and V. Niem FAO: Rome. 1106-1113 pp.

193. Nakamura, I. and N.V. Parin, (1993). FAO species catalogue. Volume 15: Snake mackerels and cutlassfishes of the world (families Gempylidae and Trichiuridae): An annotated and illustrated catalogue of the snake mackerels, snoeks, escolars, gemfishes, sackfishes, domine, oilfish, cutlassfishes, scabbardfishes, hairtails, and frostfishes known to date. *FAO Fisheries Synopsis* 125: 1-136.

194. Alexander, J., Autrup, H., Bard, D., Carere, A., Costa, L.G., Cravedi, J-P, Di Domenico, A., Fanelli, R. Fink-Gremmels, J., Gilbert, J., Grandjean, P., Johansson, N., Oskarsson, A., Ruprich, J., Schlatter,

J., Schoeters, G., Schrenk, D., van Leeuwen, R. and P. Verger, (2004). Opinion of the scientific panel on contaminants in the food chain on a request from the commission related to the toxicity of fishery products belonging to the family Gempylidae. *The European Food Safety Authority Journal* 92: 1-5.

195. Smith, P.J., R.I.C.C. Francis, and M. McVeagh, (1991). Loss of genetic diversity due to fishing pressure. *Fisheries Research* 10(3-4): 309-316.

196. Smith, P.J., L. Hauser, and G.J. Adcock (2003). Overfishing leads to loss of genetic diversity in Tasman Bay snapper. *Water & Atmosphere* 11(1): 7.

197. Fama, T., (1937). Maori Fishing. *The New Zealand Railways Magazine* 11(11): 45-46.

198. Mabin, B., (1928). Big game fishing in New Zealand's northern waters. *The New Zealand Railways Magazine* 3(5): 5-53.

199. Clarke, F.E., (1899). Notes on New Zealand Galaxidae, more especially those of the Western Slopes: with Descriptions of new species, etc. *Transactions and Proceedings of the New Zealand Institute* 31: 78-91.

200. Moss, B.G., (1958). Upokororoko, New Zealand's mystery fish. *Ammohouse Bulletin* 1(5): 1-5.

201. Furey, L., (1996). Oruarangi. The archaeology and material culture of a Hauraki pa. *Bulletin of the Auckland Institute and Museum* 17: 1-222.

202. Best, E., (1922). An eel fiesta. *New Zealand Journal of Science and Technology* 5: 108.

203. Buck, P., (1921). Maori food supplies of Lake Rotorua, with methods of obtaining them, and usages and customs appertaining thereto. *Transactions of the New Zealand Institute* 53: 432-451.

204. Best, E., (1924d). *The Utu piharau, or Lamprey-weir, as constructed on the Whanganui River*, Wellington, N.Z.: Govt. Printer. 25-30 pp.

205. Hutton, F.W., (1876). Notes on the Maori cooking places at the mouth of the Shag River. *Transactions and Proceedings of the New Zealand Institute* 8: 103-08.

206. Morison, R.J., (1990). Pacific atoll soils: chemistry, mineralogy, classification. *Atoll Research Bulletin* 339: 1-25.

207. Webb, T.H. and A.D. Wilson, (1995). A manual of land characteristics for evaluation of rural land. *Landcare Research Science Series* 10: 445-467.

208. Beattie, H., (1920). Nature lore of the southern Maori. *Transactions and Proceedings of the New Zealand Institute* 52: 53-77.

209. Newman, A.K., (1905). On a Stone-carved Ancient Wooden Image of a Maori Eel-god. *Transactions and Proceedings of the New Zealand Institute* 38: 130-134.

210. Cooke, S.J. and C.D. Suski, (2004). Are circle hooks an effective tool for conserving marine and freshwater recreational catch-and-release fisheries? *Aquatic Conservation: Marine and Freshwater Ecosystems* 14: 299-326.

211. Taylor, R.G., (2002). *Results of a bibliographic search comparing the effects of circle and 'J' hooks*: Florida Fish and Wildlife Research Institute. 13 pp.

212. Allen, M.S., (1996). Style and function in East Polynesian fish-hooks. *Antiquity* 70(267): 97-116.

213. Skinner, H.D., (1924). Archeology of Canterbury. II. Monck's Cave. *Records of the Canterbury Museum* 2: 151-162.

214. David, N. and C. Kramer, (2001). *Ethnoarchaeology in action*. Cambridge world archaeology, New York: Cambridge University Press. xxiv, 476 pp.

215. Leach, B.F., (1973). Fishing tackle in New Zealand a thousand years ago. *Kilwell Catalogue* 1973: 57-59.

216. Stewart, H., (1977). *Indian fishing: early methods on the northwest coast*, Seattle: University of Washington Press. 181 pp.

217. Powell, R., ed. (1964). Dropline fishing in deep water. [Experimental fishing for Ruvettus using modern adaptation of Polynesian techniques], *Modern Fishing Gear of the World 2*, ed. H. Kristjonsson. Fishing News Books: London. 287-291 in 578 pp.

218. Reinman, F., ed. (1970). Fishhook variability: implications for the history and distribution of fishing gear in Oceania. *Studies in Oceanic Culture History*, ed. R.K. Green, M. Vol. 11. Pacific Anthropological records. 47-59, in 176 pp.

219. Furey, L., (2002). Houhora: A fourteenth century Māori village in Northland. *Bulletin of the Auckland Institute and Museum* 19: 1-169.

220. Aerssens, J., *et al.*, (1998). Interspecies Differences in Bone Composition, Density, and Quality: Potential Implications for in Vivo Bone Research. *Endocrinology* 139(2): 663-670.

221. Reidenberg, J.S., (2007). Anatomical adaptations of aquatic mammals. *The Anatomical Record: Advances in Integrative Anatomy and Evolutionary Biology* 290(6): 507-513.

222. Zylberberg, L., *et al.*, (1998). Rostrum of a toothed whale: ultrastructural study of a very dense bone. *Bone* 23(3): 241-247.

223. Hjarno, J., (1967). Maori fish-hooks in Southern New Zealand. *Records of the Otago Museum (Anthropology)* 3: 1-63.

224. Campana, D.V., (1989). Natufian and Protoneolithic Bone Tools: The Manufacture and Use of Bone Implements in the Zagros and the Levant. *Oxford: B.A.R. BAR International Series* 494: 170.

225. Evans, F.G., (1973). *Mechanical properties of bone*, Springfield: Charles C. Thomas. 332 pp.

226. Guthrie, R.D., (1983). Osseous projectile points: biological considerations affecting raw material selection and design among Paleolithic and Paleoindian peoples, in *Animals and Archaeology. 1. Hunters and Their Prey*. p. 273-294 in J. Clutton-Brock and C. Grigson, Editors.

227. Pasveer, J.M., (2004). The djief hunters, 26,000 years of rainforest exploitation on the Bird's Head of Papua, Indonesia, *Modern Quaternary Research in Southeast Asia Vol. 17*, London: Taylor & Francis Ltd. 450 pp.

228. Knecht, H., (1997). Projectile points of bone, antler, and stone. Experimental explorations of manufacture and use, in *Projectile Technology*. p. 191-212 in H. Knecht, Editor. New York: Plenum Press

229. Paulin, C.D., (2012a). A unique Māori fish-hook – rediscovery of another Cook voyage artefact. *Tuhinga – Records of the Museum of New Zealand Te Papa Tongarewa* 23: 9-15.

230. Crosby, E.B.V., (1966). Maori Fishing Gear: A study of the development of Maori fishing gear, particularly in the North Island, in M.A. Thesis, Anthropology. University of Auckland: Auckland. p. 457.

231. Jacomb, C., (2000). Panau: The archaeology of a Banks Peninsula Māori village. *Canterbury Museum Bulletin* 9: 1-137.

232. Sinoto, Y., (1991). A revised system for the classification and coding of Hawai'ian fishhooks. *Bishop Museum Occasional Papers* 31: 1-160

233. Davidson, J. and F. Leach, (2008). A cache of one-piece fishhooks from Pohara, Takaka, New Zealand. *Terra Australis* 29: 185-202.

234. Wissler, C., ed. (1927). *Introduction. Wooden hooks used for catching sharks and Ruvettus in the South Seas: A study of their variation and distribution*, ed. E.W. Gudger. Museum of Natural History: New York. 206-208, in 150 pp.

235. Forster, G.R., (1973). Line fishing on the continental slope. The

selective effect of different hook patterns. *Journal of the Marine Biological Association of the United Kingdom* 53: 749-751.

236. Orsi, J.A., A.C. Wertheimer, and W. Jaenicke, (1993). Influence of selected hook and lure types on catch, size, and mortality of commercially troll-caught Chinook salmon. *North American Journal of Fisheries Management* 13: 709-722.

237. Hirst, K. *Ancient life in the Western Sahara desert*. February 2012; Available from: http://archaeology.about.com/od/africa/ss/gobero_4.htm.

238. Sereno, P., *et al.*, (2008). Lakeside cemetery in the Sahara: 5000 years of Holocene population and environmental change. *PLoS ONE* 2995: 3-6.

239. Herteig, A., ed. (1975). The excavation of Bryggen, Bergen, Norway. *Archaeological Contributions to the Early History of Urban Communities in Norway*, ed. A. Herteig, H.E. Lidén, and C. Blindheim. Vol. 27. Universitetsförlaget: Oslo. 174 pp.

240. Groot, G.J., (1951). *The Prehistory of Japan*, New York: Columbia University Press. 128 pp.

241. Trotter, M.M., (1956). Maori shank barbed hooks. *Journal of the Polynesian Society* 65(3): 245-252.

242. Paulin, C.D., (2012b). The traditional Māori 'internal barb' fishhook. *Tuhinga Records of the Museum of New Zealand Te Papa Tongarewa* 23: 1-8.

243. Parker, T.J. and W.A. Haswell, (1897). *A text-book of zoology*. Vol. 2, London: Macmillan and Co., limited. 2 vols., 779, 683 pp.

244. Gabriel, O., *et al.*, (2005). *Fish Catching Methods of the World*, Oxford: Blackwell Publishing Ltd 537 pp.

245. Harris, M., (1979). *Cultural Materialism: The Struggle for a Science of Culture*, New York: Random House. 381 pp.

246. Maxwell, K., (2010). One hundred years of the Otago groper fishery. *Oral Paper Abstracts New Zealand Marine Sciences Society Conference: Past, Present, Future*: 29.

247. Royal, C., (2007). *Mātauranga Māori and Museum Practice: A Discussion*. Te Papa National Services: Wellington. p. 75.

248. Fisher, V.F., (1935). The material culture of Oruarangi, Matatoki, Thames 2: Fishhooks. *Records of the Auckland Institute and Museum* 1(6): 287-300.

249. Skinner, H.D., (1943). A classification of the fish hooks of Murihiku. *Journal of the Polynesian Society* 51(3): 208-221.

250. Fairfield, F.G., (1933). Maori fish-hooks from Manakau Heads, Auckland. *The Journal of the Polynesian Society* 42(167): 145-155.

251. Anderson, A.J. and W. Gumbley, (1996). Fishing Gear, in *Shag River mouth: the archaeology of an early southern Maori village*. p. 148-160. In 294 in A.J. Anderson, Allingham, B. and I.W.G. Smith, Editor: Australian National University, Research Papers in Archaeology and Natural History.

252. Duff, R., (1942). Moa-hunters of the Wairau. *Records of the Canterbury Museum* 5: 1-42.

253. Best, E., (1974). *The stone implements of the Maori*, Wellington, N.Z.: A. R. Shearer, Govt. Printer. 445 pp.

254. Henry, T., (1928). *Ancient Tahiti*. *Bernice P Bishop Museum Bulletin* 48 3-41.

255. Heaphy, C., (1842). *Narrative of a Residence in Various Parts of New Zealand Together with a Description of the Present State of the Company's Settlements – Narrative of a Residence in Various Parts of New Zealand*, London: Smith, Elder & Co. 116 pp.

256. Lockerbie, L., (1940). Excavations at Kings Rock, Otago, with a discussion of the fish-hook barb as an ancient feature of Polynesian culture. *Journal of the Polynesian Society* 49: 393-446.

257. Teviotdale, D., (1932). The material culture of the moa-hunters in Murihiku. *Journal of the Polynesian Society* 41: 81-120.

258. Smith, I.W.G., (2007). Metal Pa Kahawai: a post-contact fishing lure form in northern New Zealand, in *Vastly Ingenious: Essays on Oceanic Material Culture*. p. 69-78 in A. Anderson, K. Green, and F. Leach, Editors. Dunedin: Otago University Press, 319pp.

259. Duff, R., (1956). *The Moa-hunter period of Maori culture*, Wellington: R.E. Owen. 400 pp.

260. Paulin, C.D., (1998). *Common New Zealand Marine Fishes*, Christchurch, N.Z.: Canterbury University Press. 80 pp.

261. Devenie, A., (2010). Pers. comm. To Phil Heatley, Minister of Fisheries. July 2010. Southern Seafoods International Ltd.

262. Hartill, B., (2009). Assessment of the KAH 1 Fishery for 2006. *New Zealand Fisheries Assessment Report* 24: 43.

263. Day, K., (2005). James Butterworth and the Old Curiosity Shop, New Plymouth, Taranaki. *Tuhinga – Records of the Museum of New Zealand Te Papa Tongarewa* 16: 93-126.

264. Loeb, E.M., (1926). History and Traditions of Niue. *Bishop Museum Bulletin* 32: 226.

265. Ryan, T.F., (1981). Fishing in Transition on Niue. *Journal de la Societe des oceanistes* 37: 193-203.

266. Leach, F., J. Davidson, and J.S. Athens, eds. (1996). Mass harvesting of fish in the waterways of Nan Madol, Pohnpei, Micronesia. *Oceanic Culture History: Essays in honour of Roger Green*, ed. J.M. Davidson, *et al.* New Zealand Journal of Archaeology Special Publication: Dunedin. 319-341, 691 pp.

267. Hooper, S., (2006). *Pacific encounters, art and divinity in Polynesia 1760-1860*, London: Te Papa Press with the British Museum 278 pp.

268. Duperrey, L.-I., *et al.*, (1828). *Voyage autour du monde execute par ordre du roi, sur la corvette de Sa Majeste La Coquille, pendant les années 1822, 1823, 1824 et 1825*. Vol. v.2 (1829), Paris: Arthus Bertrand. 164 pp.

269. Angas, G.F., (1847). *The New Zealanders Illustrated*, London: Thomas McLean. 129 pp.

270. Bagnall, L.J., (1886). Kahikatea as a building timber. *Transactions and Proceedings of the New Zealand Institute* 19: 577-580.

271. Begg, A.C. and N.C. Begg, (1979). *The world of John Boultbee*, Christchurch: Whitcoulls. 330 pp.

272. Lockerbie, L., (1959). From moa-hunter to classic Maori in Southern New Zealand, in *Anthropology in South Seas: essays presented to H.D. Skinner*. p. 75-110, 267 in J.D.a.G. Freeman, W.R., Editor. New Plymouth: Thomas Avery & Sons

273. Banks, J., (1896). *Journal of the Right Hon. Sir Joseph Banks: During Captain Cook's First Voyage in H.M.S. Endeavour in 1768-71 to Terra del Fuego, Otahite, New Zealand, Australia, the Dutch East Indies, etc*, London: McMillan & Co., 500 pp.

274. MacLaren, P.I.R., (1955). Netting knots and needles. *Man* 105: 85-89.

275. Best, E., (1934). *The Maori as he was: a brief account of Maori life as it was in pre-European days*. Manual/New Zealand Board of Science and Art, Wellington: Dominion Museum. xv, 280 p. pp.

276. Kohere, R.T., (1951). *The Autobiography of a Maori*, Wellington: Reed Publishing (NZ) Ltd. 156 pp.

277. Trewby, M., Walker, P., Schwass, M., Carew, A., Jacob, H., Bailey, S., Cormack, I. and Taylor, C. (eds), (2004). *Icons ngā taonga: from the Museum of New Zealand Te Papa Tongarewa*, Wellington: Te Papa Press. 306 pp.

278. Kaeppler, A.L., (2010). *Polynesia. The Mark and Carolyn Blackburn Collection of Polynesian Art*, Hawai'i: University of Hawai'i Press. 410 pp.

279. Kaeppler, A.L. and R. Fleck, (2009). *James Cook and the Exploration of the Pacific*, London: Thames and Hudson. 276 pp.

280. Cousins, G., (1894). *The Story of the South Seas, written for young people*, London: London Missionary Society. 248 pp.

281. Anonymous, (2010a). *Tribal Art Auction including the Webster Collection part II*, Auckland: Dunbar Sloane. 54 pp.

282. Heal, S., (2006). The great giveaway. *Museums Journal* 106(10): 32-39.

283. Waterfield, H. and J.C.H. King, eds. (2006). *Provenance: twelve collectors of ethnographic art in England 1760-1990*. Somogy Éditions d'Art: Paris. 176 pp.

284. Lowe, M., (2002). *Maori artefacts fetch top dollar*, in *Sunday Star Times*. p. 12.

285. Anonymous, (2011a). *Tribal Art Auction including the Webster Collection Part III*, Auckland: Dunbar Sloane. 80 pp.

286. Legge, C.C. and E.Q. Nash, (1969). James Edward Little, Dealer in Savage Weapons, Curios, Skins, Horns, Ivory, Etc, An Object Lesson. *Bulletin of the Field Museum of Natural History* 40(7): 9-12.

287. Paulin, C.D., (2009). Porotaka hei matau – a traditional Māori tool? *Tuhinga: Records of the Museum of New Zealand Te Papa Tongarewa* 20: 15-21.

288. Hamilton, A., (1893). Notes on Maori necklaces. *Transactions and Proceedings of the New Zealand Institute* 25: 491-493.

289. Skinner, H.D., (1916). Maori necklaces and pendants of human teeth and imitation teeth. *Man* 16: 129-130.

290. Skinner, H.D., (1933). Maori amulets in stone, bone, and shell. Part 2 & 3. *Journal of the Polynesian Society* 42: 1-9, 191-203.

291. Skinner, H.D., (1936). Maori amulets in stone, bone, and shell. Supplementary paper. *Journal of the Polynesian Society* 45: 127-141.

292. Orchiston, D.W., (1972a). Maori neck and ear ornaments of the 1770s: a study in protohistoric ethnoarchaeology. *Journal of the Royal Society of New Zealand* 2: 91-107.

293. Orchiston, D.W., (1972b). Maori greenstone pendants in the Australian Museum, Sydney. *Records of the Australian Museum* 28: 161-213.

294. Kaeppler, A.L., (1978a). *'Artificial curiosities': being an exposition of native manufactures collected on the three Pacific voyages of Captain James Cook, R.N., at the Bernice Pauahi Bishop Museum, January 18, 1978-August 31, 1978 on the occasion of the Bicentennial of the European Discovery of the Hawaiian Islands by Captain Cook – January 18, 1778*, Honolulu: Bishop Museum Press. 293 pp.

295. Golson, J., (1960). Archaeology, tradition, and myth in New Zealand prehistory. *Journal of the Polynesian Society* 69: 380-402.

296. Groube, L.M., (1967). A note on the hei-tiki. *Journal of the Polynesian Society* 76: 453-457.

297. Beck, R.J. and M. Mason, (2010). *Pounamu: the jade of New Zealand*, North Shore: Penguin in association with Ngāi Tahu. 240 pp.

298. Skinner, H.D., (1974). *Comparatively Speaking: Studies in Pacific Material Culture 1921-1972*: University of Otago Press, John McIndoe 197 pp.

299. Prickett, N., (2007). Early Maori disc pendants, in *Vastly ingenious: the archaeology of Pacific material culture*. p. 29-42 in A. Anderson, K. Green, and F. Leach., Editor. Dunedin: Otago University Press. 319 pp

300. Conly, T., (1948). Greenstone in Otago in Post-Maori times. Notes on lapidaries working in Dunedin. *Journal of the Polynesian Society* 57(1): 57-63.

301. Shortland, E., (1856). *Traditions and Superstitions of the New Zealanders: With Illustrations of their Manners and Customs*, London: Longman, Brown, Green, Longmans and Robert. 316 pp.

302. Mead, S.M., ed. (1985). Ka tupu te toi whakairo ki Aotearoa. Becoming Māori Art. *Te Māori: Māori Art from New Zealand Collections*, ed. S.M. Mead. Heinemann Ltd: Auckland. pp 63-75 in 244 pp.

303. Henare, A., (2005). *Museums, Anthropology and Imperial Exchange*, Cambridge: Cambridge University Press. 344 pp.

304. McLauchlan, G., (1981). *The Farming of New Zealand: the people and the land*, Auckland: Penguin. 264 pp.

305. Petrie, H., (2002). Colonisation and the Involution of the Māori Economy, in *XIII World Congress of Economic History*: Buenos Aires. p. 1-20.

306. Allen, G.F., (1894). *Willis's guide book of new route for tourists: Auckland-Wellington, via the hot springs, Taupo, the volcanoes and the Wanganui River*, Wanganui: A.D. Willis. 176 pp.

307. Chapman, G.T., (1881). *The natural wonders of New Zealand (the wonderland of the Pacific): its boiling lakes, steam holes, mud volcanoes, sulphur baths, medicinal springs, and burning mountains*, Auckland, New Zealand: G.T. Chapman. 172 pp.

308. Wevers, L., (2002). *Country of Writing: Writing Travel and Travel Writing About New Zealand 1809-1935*, Auckland: Auckland University Press. 234 pp.

309. Baeyertz, C.N., (1903). *Guide to New Zealand, the most wonderful scenic country in the world. The home of the Maori. The angler's and deerstalker's paradise*, Dunedin: Mills, Dick & Co. 153 pp.

310. Samson, J.O., (2003). Cultures of collecting: Māori curio collecting in Murihiku, 1865-1975: a dissertation submitted for the degree of Doctor of Philosophy at the University of Otago, Dunedin, New Zealand. University of Otago. p. xiv, 383.

311. Butterworth, J., (1901). *Catalogue of Maori Curios collected by James Butterworth of 'The Old Curiosity Shop'*, A.A. Ambridge: New Plymouth.

312. Butterworth, J., (1905). *Catalogue of Maori Curios collected by the late James Butterworth of 'The Old Curiosity Shop'*, New Plymouth: Joseph Hooker and Co. 10 pp.

313. Cowan, J., (1922). *The New Zealand wars: a history of the Maori campaigns and the pioneering period*, Wellington: W.A.G. Skinner. 280 pp.

314. Edwards, G.H., (1974). Captain John Peter Bollons. *New Zealand Marine News* 25(4): 99-118.

315. Leach, B.F., (1972). A hundred years of Otago archaeology: A critical review. *Records of the Otago Museum* 6: 1-19.

316. Reischek, A., (1930). *Yesterdays in Maoriland. New Zealand in the 'Eighties. Translated and edited by H.E.L. Priday*, London: Jonathan Cape. 311 pp.

317. King, M., (1982). *The Collector. A biography of Andreas Reischek*, Auckland: Hodder & Stoughton pp.

318. Cherry, S., (1990). *Te Ao Maori: The Maori World*, Dublin National Museum of Ireland. 56 pp.

319. Allan, T., Fleming, F. and Kerrigan, M., (1999). *Journeys through dreamtime: Oceanian myth (myth and mankind)* Amsterdam: Time-Life Books. 144 pp.

320. Bracken, T., (1879). *The New Zealand Tourist*, Dunedin: Union Steam Ship Company of New Zealand. 90 pp.

321. Scandrett, W.B., (1888). *Southland: a guide to its resources, industries and scenery by a traveller. With an interesting descriptive sketch of Stewart Island by Charles Trail, and of Invercargill*, Dunedin: Dick Mills. 32 pp.

322. Hanson, A., (1989). The Making of the Maori: Culture Invention and Its Logic. *American Anthropologist* 91(4): 890-902.

323. Hobsbawm, E.J. and T.O. Ranger, (1983). *The Invention of Tradition*, Cambridge: Cambridge University Press. 307 pp.

324. Linnekin, J.S., (1983). Defining Tradition: Variations on the Hawai'ian Identity. *American Ethnologist* 10(2): 241-252.

325. Niech, R., (1983). The Veil of Orthodoxy: Rotorua Ngati Tarawhai Woodcarving in a Changing Context. Pp 23-46 In: S.M. Mead and B. Kernot (eds.), *Art and Artists in Oceania*, Palmerston North: Dunmore. 350 pp.

326. Sissons, J., (1993). The Veil of Orthodoxy: Rotorua Ngati Tarawhai Woodcarving in a Changing Context. Pp 23-46 In: S.M. Mead and B. Kernot (eds.), Art and Artists in Oceania. *Oceania* 64(2): 1-116.

327. Sissons, J., (1998). The Traditionalisation of the Maori Meeting House. *Oceania* 69(1): 36-46.

328. Fox, J.J., ed. (1993). *Inside Austronesian Houses*. ANU E Press Australian National University: Canberra. 237 pp.

329. Groube, L.M., (1964). Settlement patterns in prehistoric New Zealand. Unpublished MA Thesis, Auckland: Auckland University. 1-290 pp.

330. van de Wijdeven, P.J.M., (2009). From art souvenir to tourist kitsch: a cultural history of New Zealand paua shell jewellery until 1981. Unpublished PhD Thesis, Department of History, Dunedin: University of Otago pp.

331. Freeman, J.D., (1949). The Polynesian collection of Trinity College, Dublin; and the National Museum of Ireland. *The Journal of the Polynesian Society* 58(1): 1-18.

332. Kaeppler, A.L., (1978b). Cook Voyage Artefacts in Leningrad, Berne and Florence Museums. *Bernice P. Bishop Museum Special Publication* 66: 168.

333. Digby, K.H., (1810-1817). *The naturalists companion containing drawings with suitable descriptions of a vast variety of quadrupeds, birds, fishes, serpents and insects; &c accurately copied either from living animals or from the stuffed specimens in the museums of the College and Dublin Society, to which is added drawings of several antiquities, natural productions & contained in those Museums, in Bound volume of watercolours*. New South Wales State Library: Sydney, Australia. p. 544.

334. Tapsell, P., (1997). The flight of Pareraututu: An investigation of taonga from a tribal perspective. *Journal of the Polynesian Society* 106: 323-374.

335. Māhina-Tuai, K.U., (2006). Intangible heritage: A Pacific case study at the Museum of New Zealand Te Papa Tongarewa. *International Journal of Intangible Heritage* 1: 14-24.

336. Hector, J., (1870). *Catalogue of the Colonial Museum, Wellington, New Zealand*, Wellington: Government Printer. 56 pp.

337. Smith, W.J., (1965). Sir Ashton Lever of Alkrington and his museum, 1729-88 *Transactions of the Lancashire & Cheshire Antiquarian Society* 72: 61-92.

338. Roberts, C.D. and C.D. Paulin, (1997). Fish collections and collecting in New Zealand, in *Collection building in ichthyology and herpetology*. p. 201-229 in T.W. Pietsch, W.D. Jr. (eds), Editor: American Society of Ichthyologists and Herpetologists Special Publication

339. Nathan, S. and M. Varnham, eds. (2008). *The amazing world of James Hector: explorer, geologist, botanist, natural historian, surgeon – and one of New Zealand science's most remarkable figures*. Awa Press: Wellington. 183 pp.

340. Anonymous. *The founding and building of the Auckland museum*. 2010b [cited 2010; Available from: http://www.aucklandmuseum. com/159/history-of-the-museum.

341. Hector, J., (1869). Preface. *Transactions and Proceedings of the New Zealand Institute* 1: i-iv.

342. Shawcross, W., (1970). The Cambridge University collection of Maori artefacts, made on Captain Cook's "first voyage". *Journal of the Polynesian Society* 79(3): 305-348.

343. Shawcross, W., (1970). The Cambridge University collection of Maori artifacts, made on Captain Cook's "first voyage". *Journal of the Polynesian Society* 79 (3): 305-348.

344. Gathercole, P., (1991). *The Maori collection at the Cambridge University Museum of Archaeology and Anthropology. Papers of taonga Maori conference New Zealand, 18-27 November 1990*, Wellington: Cultural Conservation Advisory Council/Te Roopu Manaakii Nga Taonga Tuku Iho and Department of Internal Affairs, Te Tari Taiwhenua. 179 pp.

345. Kaeppler, A.L., (2008). *To attempt some new discoveries in that vast unknown tract*, in *Cook's Pacific Encounters symposium*: National Museum of Australia.

346. Frondel, C., (1972). Jacob Forster (1739-1806) and his connections with forsterite and palladium, *Mineralogical Magazine* 38: 545-550.

347. Whitehead, P.J.P., (1973). Some further notes on Jacob Forster (1739-1806), mineral collector and dealer. *Mineralogical Magazine* 39: 361-363.

348. Kruger, G., (2009). *Institut für Ethnologie der Universität Göttingen*: pers. comm.

349. Paulin, C.D., (in prep). A unique Māori fish-hook – rediscovery of another Cook voyage artefact. *Tuhinga – Records of the Museum of New Zealand Te Papa Tongarewa*.

350. Donovan, E., (1806). *Catalogue of the Leverian Museum*. Vol. Parts I-VII + Appendix, London. 348 pp.

351. Gores, S.J., (2000). *Psychosocial spaces: verbal and visual readings of British culture 1750-1820*, Detroit: Wayne State University Press. 223 pp.

352. King, J.C.H., (1996). New evidence for the contents of the Leverian Museum. *Journal of the History of Collections* 8(2): 167-186.

353. Phillips, W.J., (1930). Maori burial chests in Vienna Museum. *Journal of the Polynesian Society* 39(156): 388-389.

354. Hooper, S., (2007). in *Catalog Notes*, H. Museum, Editor.

355. Coote, J., (2000). Curiosities sent to Oxford: The original documentation of the Forster collection at the Pitt Rivers Museum. *Journal of the History of Collections* 12(2): 177-192.

356. Coote, J., (2004). *Curiosities from the Endeavour: A forgotten collection – Pacific artefacts given by Joseph Banks to Christ Church, Oxford after the first voyage*, Whitby: Captain Cook Memorial Museum. 28 pp.

357. Beechey, F.W., (1831). *Narrative of a voyage to the Pacific and Beering's Strait, to co-operate with the Polar Expeditions: performed in His Majesty's Ship Blossom under the Command of Captain F.W. Beechey, R.N., F.R.S. &c., in the Years 1825, 26, 27, 28*, London Henry Colburn and Richard Bentley. 452 pp.

358. Coote, J., (2012). Objects and words: writing on, around, and about things – an introduction. *Journal of Museum Ethnology* 25: 3-18.

359. Henking, K.H., (1957). Die Südsee – und Alaskasammlung Reinhold Wäber. Beschreibender Katalog. *Jahrbuch der Bernischen Historischen Museums in Bern* 35/36: 325-389.

360. Giglioli, E.H., (1893). *Appunti intorno ad una collezione etnografica fatta durante il terzo viaggio di Cook, e conservata sin dalla fine del secolo scorso nel R. Museo di Fisica e Storia Naturale di Firenze* Florence: Tipografia di Salvadore Landi. 161 pp.

361. Campbell, N. and J. Hobbs, (2012). *Important Oceanic & African Art*, in *Auction Catalogue*, Webb's, Editor: Auckland.

362. Anderson, A. and M. McGlone, eds. (1991). *Living on the edge – prehistoric land and people in New Zealand*, pp. 199-241 in The Naive Lands: Prehistory and Environmental change in Australia and the South-west Pacific, ed. J. Dodson. Longman Cheshire: Melbourne. 386 pp.

363. Nichol, R., (1988). Tipping the feather against a scale: archaeozoology from the tail of the fish, in *Department of Anthropology*. University of Auckland. p. 1-572.

364. Francis, M.P., (1993). Does water temperature determine year class strength in New Zealand snapper (Pagrus auratus, Sparidae)? *Fisheries Oceanography* 2(2): 65-72.

365. Leach, B.F., *et al.*, (1999). Pre-European catches of barracouta, *Thyristes atun*, at Long Beach and Shag River Mouth, Otago, New Zealand. *Archaeofauna* 8: 11-30.

366. Leach, B.F., J. Davidson, and K. Fraser, (2000). Pre-European catches of blue cod (*Parapercis colias*) in the Chatham Islands and Cook Strait, New Zealand. *New Zealand Journal of Archaeology* 21: 119-138.

367. Vogel, J., (1875). *The official handbook of New Zealand. A collection of papers by experienced colonists on the colony as a whole, and on the several provinces*, London: Wyman & Sons. 272 pp.

368. Pearson, W.H., (1872). Stewart's Island as a field for settlement. *Otago Witness* 13 July 1872: 8.

369. Hector, J., (1872). *Notes on the Edible Fishes of New Zealand*, Wellington: J. Hughes. 95-133 pp.

370. Ayson, L.F., (1913). Fisheries of New Zealand. *Appendix to the Journal of the House of Representatives* H. (-15b.): 1-18.

371. Baden, J.A. and D.S. Noonan, (1998). *Managing the Commons*, Indiana: Indiana University Press. 264 pp.

372. Hector, J., (1897). Protection of mullet. *Appendix to the Journal of the House of Representatives* H-17: 1-24.

373. Memon, P. and R. Cullen, (1992). Fishery Policies and their Impact on the New Zealand Maori. *Marine Resource Economics*. VII (3): 153-67.

374. McKenzie, J., (1885). *Development of colonial industries. Paper presented to both Houses of the General Assembly.*, Wellington: Government Printer. 9 pp.

375. Mackenzie, J., ed. (1885). The Hon. Sir Julius von Haast. *Handbook of the Fishes of New Zealand*, ed. R.A.A. Sherrin. Wilson and Horton: Auckland. pp. 105-108, in 307 pp.

376. Barlow, P.W., (1888). *Kaipara, or experiences of a settler in north New Zealand*, London: Marston, Searle and Revington. 219 pp.

377. Paulin, R., (1892). The Red Hill and Awaruiti District. *Otago Daily Times* 20th February 9355: 1.

378. May, P.R., (1962). *The West Coast Gold Rushes*, Christchurch: Pegasus. 559 pp.

379. Great Britain Hydrographic Office, (1946). *The New Zealand pilot: comprising the coasts of the North and South Islands of New Zealand, Stewart Island and adjacent islands, Kermadec, Chatham, Bounty, Antipodes, Auckland and Campbell Islands*, London: Printed for the Hydrographic Office, Admiralty by H.M.S.O. 447 pp.

380. Maling, P.B., (1969). *Early charts of New Zealand, 1542-1851*, Wellington: A.H. & A.W. Reed. 134 pp.

381. Thomson, G.M., (1896). New Zealand fisheries and the desirability of introducing new species of sea fish. *Proceedings of the New Zealand Institute* 28: 758.

382. Thomson, G.M. and T. Anderton, (1921). *History of the Portobello marine fish-hatchery and biological station*, Wellington: Government Printer. 131 pp.

383. Thorn, J., Young, M.W., and Sheed, E., (1938). *Report of the sea fisheries investigation committee*. Vol. H.-44A., Wellington: Appendix to the Journal of the House of Representatives 128 pp.

384. Gibbs, M.T., (2008). The historical development of fisheries in New Zealand with respect to sustainable development principles. *Electronic Journal of Sustainable Development* 1 (2): 23-33.

385. Stewart, C., (2004). *Legislating for property rights in fisheries*, Rome: Food and Agriculture Organization of the United Nations. 201 pp.

386. Thomson, P., (1876). Fish and their Seasons. *Transactions and Proceedings of the New Zealand Institute* 9: 484-490.

387. Thomson, P., (1877). The Dunedin fish supply. *Transactions and Proceedings of the New Zealand Institute* 10: 324-330.

388. Anderton, T., (1906). Observations on New Zealand Fishes, &c., made at the Portobello Marine Fish-hatchery. *Transactions and Proceedings of the New Zealand Institute* 39: 477-495.

389. Graham, D.H., (1938a). Food of the Fishes of Otago Harbour and adjacent sea. *Transactions and Proceedings of the New Zealand Institute* 68: 421-436.

390. Graham, D.H., (1938b). Fishes of Otago Harbour and adjacent seas with additions to previous records. *Transactions and Proceedings of the New Zealand Institute* 68: 399-419.

391. Thomson, G.M., (1906). The Portobello Marine Fish Hatchery and Biological Station. *Transactions and Proceedings of the New Zealand Institute* 38: 529-558.

392. Thomson, G.M., (1912). The Natural History of Otago Harbour and the Adjacent Sea, together with a Record of the Researches carried on at the Portobello Marine Fish-hatchery: Part I. *Transactions and Proceedings of the New Zealand Institute* 45: 225-251.

393. Ministry of Fisheries, (2010). East Otago Taiapure – Regulatory Interventions. *Regulatory Impact Statement*: 5.

394. Grey, Z., (1926). *Tales of the Angler's El Dorado, New Zealand*, London: Hodder & Stoughton. 228 pp.

395. D'Esterre, D., (1907). In Auckland: happy hunting ground of the deep sea angler, in *Otago Witness*: Dunedin.

396. Graham, D.H., (1953). *A Treasury of New Zealand Fishes*, Wellington: A.H. & A.W. Reed. 404 pp.

397. Hughey, K.F.D., (1997). Fisheries Management in New Zealand – privatising the policy net to sustain the catch? *Environmental Politics* 6: 4.

398. Jung, C.A., *et al.*, (2010). Perceptions of environmental change over more than six decades in two groups of people interacting with the environment of Port Phillip Bay, Australia. *Ocean & Coastal Management* 54(1): 93-99.

399. Koenig, C.C. and F.C. Coleman, (2009). Population density, demographics, and predation effects of adult goliath grouper (*Epinephelus itajara*). *Final Report to NOAA MARFIN for Project NA05NMF4540045*: 79.

400. McClenachan, L., (2009). Historical declines of goliath grouper (*Epinephelus itajara*) populations of South Florida, USA. *Endangered Species Research* 7: 175-181.

401. Parsons, D.M., *et al.*, (2009). Risks of shifting baselines highlighted by anecdotal accounts of New Zealand's snapper (*Pagrus auratus*) fishery. *New Zealand Journal of Marine and Freshwater Research*: 965-983.

402. Fisheries, M.o., (1984). *Inshore finfish fisheries: proposed policy for future management*. New Zealand Ministry of Fisheries. p. 31.

403. Habib, G., (1989). *Report on Ngaitahu Fisheries Evidence*, in

Ngaitahu Claim to Mahinga Kai, W. Tribunal, Editor., Department of Justice: Auckland. p. 363.

404. Clement, G., (2009). *New Zealand commercial fisheries: The atlas of area codes and TACCs 2009/2010*, Nelson: Clement and Associates Ltd. 90 pp.

405. Statistics New Zealand, (2010). *Fish monetary stock account 1996-2009* Wellington: Statistics New Zealand 53 pp.

406. Mullon, C., P. Freon, and P. Cury, (2005). The dynamics of collapses in world fisheries. *Fish and Fisheries* 6(2): 111-120.

407. McCoy, J., (2010). In Churchouse, N., Fishing limits decisions 'guesswork'. *The Dominion Post* (9 March, C1).

408. Heatley, P., (2011). Speech to the Seafood Industry Council Conference, F.a. Aquaculture, Editor: Wellington.

409. Annala, J., (1996). New Zealand's ITQ system: have the first eight years been a success or a failure? *Reviews in Fish Biology and Fisheries* 6(1): 43-62.

410. MFish, (2006). *The State of our Fisheries*, Wellington: New Zealand Ministry of Fisheries. 48 pp.

411. Hartill, B., (2004). Characterisation of the commercial flatfish, grey mullet, and rig fisheries in the Kaipara Harbour. *New Zealand Fisheries Assessment Report* 2004/1: 23 pp.

412. Field, J.C., (2002). *A review of the theory, application and potential ecological consequences of F40% harvest policies in the Northeast Pacific*, Seattle: University of Washington, School of Aquatic and Fisheries Sciences. 101 pp.

413. Goodman, D., Mangel, M., Parkes, G., Quinn, T., Restrepo, V., Smith, T. and Stokes, K., (2002). *Scientific review of the harvest strategy currently used in the BSAI and GOA groundfish fishery management plans*, Anchorage: North Pacific Fishery Management Council. 153 pp.

414. Jackson, J.B.C., *et al.*, (2001). Historical Overfishing and the Recent Collapse of Coastal Ecosystems, *Science* 293(5530): 629-637.

415. Myers, R.A., and Worm, B., (2003). Rapid worldwide depletion of predatory fish communities. *Nature* 423: 280-283.

416. Barber, W.E., (1988). Maximum sustainable yield lives on. *North American Journal of Fisheries Management* 8(2): 153-157.

417. Davis, G., (1989). Designated harvest refugia: The next stage of marine fishery management in California. *California Cooperative Oceanic Fisheries Investigations Report* 30: 1-6.

418. Larkin, P.A., (1977). An epitaph for the concept of maximum sustained yield. *Transactions of the American Fisheries Society*. 106 (1): 1-11.

419. Roedel, P.M., (ed.) (1975). Optimum sustainable yield as a concept in fisheries management. *American Fisheries Society Special Publication* 9: 1-89.

420. Sherman, D.J., (2006). Seizing the cultural and political moment and catching fish: Political development of Māori in New Zealand the Sealord Fisheries Settlement, and social movement theory. *Social Science Journal* 43(4): 513-527.

421. Struhsaker, T.T., (1998). A biologist's perspective on the role of sustainable harvest in conservation. *Conservation Biology* 12(4): 930-932.

422. Ostrom, E., *et al.*, (2002). *Drama of the Commons*, Washington, D.C.: National Academies Press. 521 pp.

423. Willis, T.J. and R.B. Millar, (2005). Using marine reserves to estimate fishing mortality. *Ecology Letters* 8: 47-52.

424. Roberts, C.M., *et al.*, (2001). Effects of marine reserves on adjacent fisheries. *Science* 294: 1920-1923.

425. Harrison, Hugo B., *et al.*, (2012). Larval Export from Marine Reserves and the Recruitment Benefit for Fish and Fisheries. *Current biology*: CB 24(11): 1023-1028.

426. Hillborn, R., (2002). Marine reserves and fishery management. *Science* 295: 1233-1234.

427. Hillborn, R., (2010). Apocalypse Forestalled: Why All the World's Fisheries Aren't Collapsing. *The Science Chronicles* (Published by The Nature Conservancy) November 2010: 5-9.

428. Roberts, C.M., *et al.*, (2001). Effects of marine reserves on adjacent fisheries. *Science* 294: 1920-1923.

429. De Groot, S.J., (1984). The impact of bottom trawling on benthic fauna of the North Sea. *Fisheries Research* 5: 39-54.

430. Jones, J.B., (1992). Environmental impact of trawling on the seabed: a review. *New Zealand Journal of Marine and Freshwater Research* 26: 59-67.

431. Seafood Industry Council. *Benthic Protection Areas* 2007 [cited 2010]; Available from: http://www.seafoodindustry.co.nz/bpa.

432. Anonymous, (2007). New Zealand designates network of deepsea protected areas covering more than one million square kilometers. *MPA News* 9(5): 1-2.

433. Stokes, K., (2006). New Zealand Seafood Industry proposes huge closures – Cynicism or Pragmatism? *MPA News* 7(9): 1.

434. Bruce, B.D., Green, M.A. and Last, P.R., (1998). Threatened fishes of the world: spotted handfish, *Brachionichthys hirsutus* (Lacepede). *Environmental Biology of Fishes* 52: 418.

435. Baum, J.K., Myers, R.A., Kehler, D.G., Worm, B., Harley, S.J. and Doherty, P.A., (2003). Collapse and conservation of shark populations in the northwest Atlantic. *Science* 299: 389-392.

436. Musick J.A., H.M.M., Berkeley S.A., Burgess G.H., Eklund A.M., Findley L., Gilmore R.G., Golden J.T., Ha D.S., Huntsman G.R., McGovern J.C., Parker S.J., Poss S.G., Sala E., Schmidt T.W., Sedberry G.R., Weeks H. and Wright S.G., (2000). Marine, estuarine, and diadromous fish stocks at risk of extinction in North America (exclusive of Pacific salmonids). *Fisheries* 25: 6-30

437. Musick, J.A., (1999). Criteria to define extinction risk in marine fishes. *Fisheries* 24: 6-14.

438. Roberts, C.M. and J.P. Hawkins, (1999). Extinction risk in the sea. *Trends in Ecology and Evolution* 14: 241-246

439. Casey, J.M. and R.A. Myers, (1998). Near extinction of a large widely distributed fish. *Science* 31(281): 690-692.

440. Gordon, H.S., (1954). The economic theory of a common-property resource: The Fishery. *Journal of Political Economy* 62: 124-42.

441. Scott, A.D., (1955). The Fishery: The objectives of sole ownership. *Journal of Political Economy* 63: 116-24.

442. FAO. *The state of world fisheries and aquaculture 2000*. 2000 [cited 2010 23rd March]; Available from: www.FAO.org.

443. Newell, R.G. *Maximising value in multi-species fisheries*. Ian Axford Fellowship in Public Policy 2004 [cited 2010 23rd March]; Available from: www.Fullbright.org.nz.

444. Branch, T.A., *et al.*, (2011). Contrasting Global Trends in Marine Fishery Status Obtained from Catches and from Stock Assessments. *Conservation Biology* 25: 1523-1739.

445. Pauly, D., (2007). The Sea Around Us Project: Documenting and communicating global fisheries impacts on marine ecosystems. *Ambio* 36: 290-295.

446. Richardson, A.J., *et al.*, (2009). The jellyfish joyride: causes, consequences and management responses to a more gelatinous future. *Trends in Ecology and Evolution* 24(6): 312-322.

447. Worm, B., *et al.*, (2006). Impacts of Biodiversity Loss on Ocean Ecosystem Services. *Science* 314: 1-169.

448. Parsons, T.R. and C.M. Lalli, (2003). Jellyfish population explosions: revisiting a hypothesis of possible causes. *La Mer* 40: 111-121.

449. Worm, B., *et al.*, (2009). Rebuilding Global Fisheries. *Science* 325: 578-585.

450. FAO, (2009). *The State of the World's Fisheries and Aquaculture 2008*, Rome: FAO. 176 pp.

451. Ministry of Fisheries, (2005). *Strategy for Managing the Environmental Effects of Fishing.* p. 20.

452. Berrill, M., (1997). *The plundered seas: can the world's fish be saved?*, Vancouver: Greystone Books. 208 pp.

453. Mason, F., (2002). The Newfoundland cod stock collapse: A review and analysis of social factors. *Electronic Green Journal* 17: Retrieved from: http://escholarship.org/uc/item/19p7z78s.

454. Cawthron, I., (2007). *Fishing in the dark. Science, values and deep water fisheries research.*, in *Master Environmental Studies.* Victoria University: Wellington. p. 131.

455. Pauly, D., *et al.*, (1998). Fishing Down Marine Food Webs. *Science* 279: 860-863.

456. Robertson, D., (2010). *Marine Biodiversity: Taxonomists dream; resource mangers nightmare? Past, Present, Future*, Wellington: New Zealand Marine Sciences Society Conference. 29 pp.

457. DeVries, A.L., Ainley D.G., Ballard G., (2008). Decline of the Antarctic toothfish and its predators in McMurdo Sound and the southern Ross Sea and recommendations for restoration. *CCAMLR document WG-EMM 08/xx.*

458. Hickling, C.F., (1975). *Water as a productive environment*, New York: St Martin's Press. 203 pp.

459. Mace, P., (2010). *Defining 'Endangered': are marine species 'Special'? Past, Present, Future*, Wellington: New Zealand Marine Science Society Conference. 29 pp.

460. Daniell, S., (1992). *Sustainable management of the Chatham Rise orange roughy fishery*, Wellington: Office of the Parliamentary Commissioner for the Environment. 21 pp.

461. Martin, A., (2010). Orange roughy fishery collapse continues. *Dominion Post*: Wellington.

462. Clark, M., (2001). Are deepwater fisheries sustainable? – the example of orange roughy (*Hoplostethus atlanticus*) in New Zealand. *Fisheries Research* 51: 123-135.

463. Norse, E.A., *et al.*, (2012). Sustainability of deep-sea fisheries. *Marine Policy* 36: 307-320.

464. Pogonoski, J., D. Hoese, and S. Reader, (2011). *Survey guidelines for Australia's threatened fish*. Department of Sustainability, Environment, Water, Population and Communities, Australian Government. p. 57.

465. Bradford-Grieve, J., *et al.*, (2007). Ocean variability and declining hoki stocks: an hypothesis as yet untested. *New Zealand Science Review* 63(3-4): 76-80.

466. Anonymous. *Marine Stewardship Council.* 2010d [cited 2010; Available from: http://www.msc.org/documents/msc-brochures/MSC-FisheriesCommitments-Aug09-WEB.pdf/view.

467. Council, M.S. *New Zealand Hoki Fishery certified to Marine Stewardship Council standard.* 2001 [cited 2011; Available from: http://www.msc.org/newsroom/news/new-zealand-hoki-fishery-certified-to-marine.

468. Ministry for the Environment, (2010). Fishing Activity: Fish Stocks, in *Environmental Report Card.* p. 7.

469. Hauser, L. and G.R. Carvalho, (2008). Paradigm shifts in marine fisheries genetics: ugly hypotheses slain by beautiful facts. *Fish and Fisheries* 9(4): 333-362.

470. Anonymous, (2009). *Updated Status of New Zealand's Fishstocks*: Ministry of Fisheries 2 pp.

471. Coggan, F., (1997). Maori Activists and Fishermen Discuss how to Protect Fish Stocks in New Zealand. *Militant* 61(32): 1.

472. Francis, R., D. Gilbert and J. Annala (1993). Fisheries management by individual quotas: theory and practice. *Marine Policy* January 1993: 64-65.

473. Morgan, G.R., (1997). *Individual Quota Management in Fisheries – Methodologies for Determining Catch Quotas and Initial Quota Allocation. FAO Fisheries Technical Paper.* 41 pp.

474. McClintock, W.L., Baines, J.T. and Taylor, C.N., (2000). Retreat From the Frontier: Fishing Communities in New Zealand, in *8th International Symposium on Society and Resource Management*: Western Washington University, Bellingham, Washington, USA. p. 1-13.

475. Coburn, R.P. and I.J. Doonan, (1994). *Orange roughy on the northeast Chatham Rise: a description of the commercial fishery, 1979-88. New Zealand Fisheries Technical Report.* Vol. 38, Wellington: New Zealand Ministry of Fisheries. 49 pp.

476. Blanchette, K., (2009). New Zealand's commercial fishing industry: too many fish in the sea. *Perspectives on Business and Economics* 27: 15-24.

477. Hauser, L., *et al.*, (2002). Loss of microsatellite diversity and low effective population size in an overexploited population of New Zealand snapper (*Pagrus auratus*). *Proceedings of the National Academy of Sciences of the United States of America* 99(18): 11742-11747.

478. Anonymous. *New Zealand albacore tuna fishery gains MSC certification.* 2011b; Available from: http://www.msc.org/newsroom/news/new-zealand-albacore-tuna-fishery-gains-msc-certification.

479. Smith, L. *Sustainable fish customers 'duped' by Marine Stewardship Council.* 2011 [cited 2012 6th May]; Available from: http://www.guardian.co.uk/environment/2011/jan/06/fish-marine-stewardship-council.

480. Anonymous. *Experts: Marine Stewardship Council's Certification Failing Basic Science; 'Bureaucracy' Instead Of 'Biology'.* 2010e [cited 2012 6th May]; Available from: http://www.underwatertimes.com/news.php?article_id=03249681051.

481. Alder, J., *et al.*, (2010). Aggregate performance in managing marine ecosystems of 53 maritime countries. *Marine Policy* 34(3): 468-476.

482. Pitcher, T.J., *et al.*, (2009). An evaluation of progress in implementing ecosystem-based management of fisheries in 33 countries. *Marine Policy* 33(223-232).

483. Mora, C., *et al.*, (2009). Management effectiveness of the world's marine fisheries. *PLoS Biology* 7(e1000131).

484. Beddington, J.R., D.J. Agnew, and C.W. Clark, (2007). Current Problems in the Management of Marine Fisheries. *Science* 316: 1713-1716.

485. Day, A., (2004). Fisheries in New Zealand: The Maori and the Quota Management System. *Paper Prepared For The First Nation Panel on Fisheries*: 12 pp.

486. Allibone, R., *et al.*, (2010). Conservation status of New Zealand Freshwater fish. *Journal of Marine and Freshwater Research* 44(4): 271-287.

487. Joy, M., (2010). Government failures and ecological apathy bite back, in *Dominion Post*. 13th August: Wellington. p. B7.

488. Doole, G.J., (2005). Optimal management of the New Zealand longfin eel (*Anguilla dieffenbachii*). *Australian Journal of Agricultural and Resource Economics* 49: 395-411.

489. Jellyman, D.J., *et al.*, (2000). A review of the evidence for a decline

in the abundance of longfinned eels (*Anguilla dieffenbachii*) in New Zealand Vol. No. EEL9802, Wellington: Ministry of Fisheries. 76 pp.

490. Neville, A., (2010). Endangered eel on sale, in *New Zealand Herald*. 10 January: Auckland

491. Clarke, S., (1989). *The First Seven Fleets*, Wellington: Wilton Library Services. 11 pp.

492. Belich, J., (1986). *The New Zealand Wars and the Victorian Interpretation of Racial Conflict*, Auckland: Auckland University Press. 396 pp.

493. Crosby, R.D., (1999). *The Musket Wars: a history of inter-iwi conflict, 1806-45*, Auckland: Reed. 392 pp.

494. Ballara, A., (2003). *Taua: 'musket wars', 'land wars' or tikanga?: warfare in Māori society in the early nineteenth century*, Auckland: Penguin. 543 pp.

495. Shortland, E., (1851). *The Southern Districts of New Zealand: a Journal, with Passing Notices of the Customs of the Aborigines*, London: Longman, Brown, Green and Longmans. 316 pp.

496. Great Britain Parliament House of Commons, (1845). *A corrected report of the debate in the House of Commons, on the 17th, 18th, and 19th of June, on the state of New Zealand and the case of the New Zealand Company*, London: J. Murray. 287 pp.

497. Heritage, M.f.C.a. *Declaration of Independence – background to the Treaty*. 2009 [cited 2011]; Available from: http://www.nzhistory.net.nz/politics/treaty/background-to-the-treaty/declaration-of-independence.

498. Orange, C., (1990). *The Story of a Treaty*, Wellington: Bridget Williams books. 80 pp.

499. Stearns, P.N. and W.L. Langer, (2001). *The Encyclopedia of World History: ancient, medieval and modern chronologically arranged*, New York: Houghton Mifflen Co. 1,243 pp.

500. Hayward, J., (1997). The Principles of the Treaty of Waitangi, in *National Overview: Waitangi Tribunal Rangahaua Whanui Series*. p. 475-494 in A. Ward, Editor: Waitangi Tribunal

501. Morris, G., (2004). James Prendergast and the Treaty of Waitangi: Judicial Attitudes to the Treaty During the Latter Half of the Nineteenth Century *Victoria University of Wellington Law Review* 35(1): 117.

502. McHugh, P.G., (1984). The legal status of Maori fishing rights in tribal waters. *Victoria University Wellington Law Review* 14: 247-258.

503. Palmer, M.S., (2008). *The Treaty of Waitangi in New Zealand's Law and Constitution*. Victoria University Press. 477 pp.

504. New Zealand House of Representatives, (1877). *Parliamentary debates*: s.n. pp.

505. Mackay, J., (1862). Reports on the state of the Natives in various Districts, at the time of the arrival of Sir George Grey. *Appendix to the Journals of the House of Representatives* (Session I, E-07): 1-42.

506. Bennoin, T. and G. Melvin. *Taranaki Fish & Game Council v McCritchie*. The Maori Law Review. A monthly review of law affecting Maori 1997; Available from: http://www.bennion.co.nz/mlr/1997/feb.html.

507. Anonymous., (1895). Marine Department annual report for 1894-1895. *Appendix to the Journals of the House of Representatives* H-29: 1-23.

508. Lock, K. and S. Leslie, (2007). *New Zealand's Quota Management System: A History of the First 20 Years (April 2007). Motu Working Paper No. 07-02*: Ministry of Fisheries. 75 pp.

509. Doig, S., (1997). The New Zealand judiciary and Māori fishing rights, 1880s/90s & 1980s/90s, in *New Zealand Historical Association Conference*: Massey University, 5 December 1997.

510. Levine, H.B., (1997). *Constructing collective identity: A comparative analysis of New Zealand Jews, Māori and Urban Papua New Guineans*, New York: Peter Lang. 179 pp.

511. Clement, I.T., T.H.A. Spear, and C.D. Paulin, (2003). *New Zealand Commercial Fisheries: The Guide to the Quota Management System*, Nelson: Clement & Associates Ltd. 92 pp.

512. Waitangi Tribunal, (1983). Report of the Waitangi Tribunal on the Motunui-Waitara Claim. *Wai* 6: 73.

513. Williams, D., (2001). *Matauranga Maori and Taonga. The nature and extent of Treaty Rights held by iwi and hapu in Indigenous flora and fauna, Cultural heritage Objects, Valued traditional knowledge*. Waitangi Tribunal: Wellington. p. 167.

514. Sissons, J., (2004). Māori Tribalism and Post-Settler Nationhood in New Zealand. *Oceania* 75: 19-31.

515. New Zealand Government, (1992). Treaty of Waitangi Fisheries Deed of Settlement. *http://www.tokm.co.nz/profiles/profiles/DeedofSettlement.pdf*.

516. Hepburn, C., (2011). Pathways to fisheries restoration through customary fisheries tools, in *Understanding, Managing and Conserving our marine Environment*, N.Z.M.S. Society, Editor: Oban, Stewart Island. p. 32.

517. Anonymous, (2006). *The State of Our Fisheries*, Wellington: Ministry of Fisheries. 48 pp.

518. Barcham, M., (1998). The challenge of urban Maori: Reconciling conceptions of indigeneity and social change. *Asia Pacific Viewpoint* 39: 303-314.

519. Bourassa, S.C., and Strong, A. L., (2000). Restitution of fishing rights to Māori: Representation, social justice and community development. *Asia Pacific Viewpoint* 41(2): 155-175.

520. McClean, R. and T. Smith, (2001). *The Crown Flora and Fauna: Legislation, Policies and Practices 1983-98*. Waitangi Tribunal Publication, Wellington. 733 pp.

521. Annala, J., Clark, M., Clement, G. and J. Cornelius (2005). Management of New Zealand orange roughy fisheries: a deep learning curve. pp. 544-554 *in*: Shotton, R. (ed.) *Deep Sea 2003: Conference on the Governance and Management of Deep-sea Fisheries*, Queenstown (New Zealand), 1-5 Dec 2003. FAO, Rome (Italy). Fishery Resources Division.

Rārangi whakaatu: *index*

Page numbers shown in **bold** in this Index indicate an illustration featuring the indexed subject in the image, or referenced in the caption.

About the authors

Chris Paulin is a self-employed marine biologist and publisher. His research on the taxonomy and systematics of fishes of the New Zealand region has been published in more than 60 scientific papers in national and international journals, as well as in numerous popular articles. His books and identification guides include *New Zealand Fish, a complete guide* (1989), *Rockpool Fishes of New Zealand* (1992), *Common New Zealand Fishes* (1998), and *New Zealand Commercial Fisheries: An Identification Guide to Quota Management Species* (1996, 1998, 2003). With Paddy Ryan he authored *Fiordland Underwater: New Zealand's Hidden Wilderness* (1998), *The Rocky Shore: a guide to the intertidal plants and animals of Wellington's Taputeranga Marine Reserve* (2014), and *Taranaki's Rocky Shore: a guide to the intertidal plants and animals* (2014). He was a recipient of the 1996 Royal Society of New Zealand Science Communicator Award. Using his knowledge of the ecology of New Zealand fishes, Chris studied Māori fish-hooks in museum collections, revealing that their unique designs and function represented an innovation in hook technology that has only recently been adopted by present day fishers. In 2009 he received a Winston Churchill Memorial Trust Fellowship that enabled him to travel to Europe to examine pre-contact Māori hooks collected by James Cook and other explorers in the 18th century.

Mark Fenwick graduated from Victoria University of Wellington with a Master's degree in Marine Biology. He has worked at the National Museum Te Papa Tongarewa in the Natural Environment collections, and at the National Institute of Water and Atmospheric Research as a marine ecology technician. Mark is descended from European and Māori ancestors. He affiliates to Te Atiawa and Taranaki iwi and is a trustee of the Te Atiawa ki Te Upoko o te Ika a Māui Potiki Trust representing Te Tatau o Te Po, his home marae in Petone. Mark has a lifelong interest in the ocean, is an active scuba instructor, diver, fisherman and environmentalist. He has published papers on freshwater mussel genetics and taxonomy and has written and co-written numerous reports, popular articles, and book chapters. He completed his thesis on the genetics of the New Zealand freshwater mussels at Victoria University.

Fish-hooks and other museum collection items photographed by the author for this publication, with permission of those institutions, include Fig.19, 21, 26, 27, 32, 34, 36, 37, 41, 42, 45, 47, 65, 69, 71, 72, 73, 75, 77, 83, 85, 88, 89, 91. These images are © Chris Paulin.